MARIANO TACCOLA
AND HIS BOOK
DE INGENEIS

The MIT Press
Cambridge, Massachusetts,
and London, England

Frank D. Prager
and Gustina Scaglia

MARIANO TACCOLA
AND HIS BOOK
DE INGENEIS

Copyright © 1972 by
The Massachusetts Institute of Technology

This book was designed by The MIT Press
Design Department.
It was set in Linofilm Palatino
by Book Graphics, Inc.
printed on Warrens Olde Style Offset
by The Colonial Press, Inc.
and bound in Lindenmeyer Elephant Hide
by The Colonial Press, Inc.
in the United States of America.

ISBN 0 262 16 045 5 (hardcover)

Library of Congress catalog card number: 70–118352

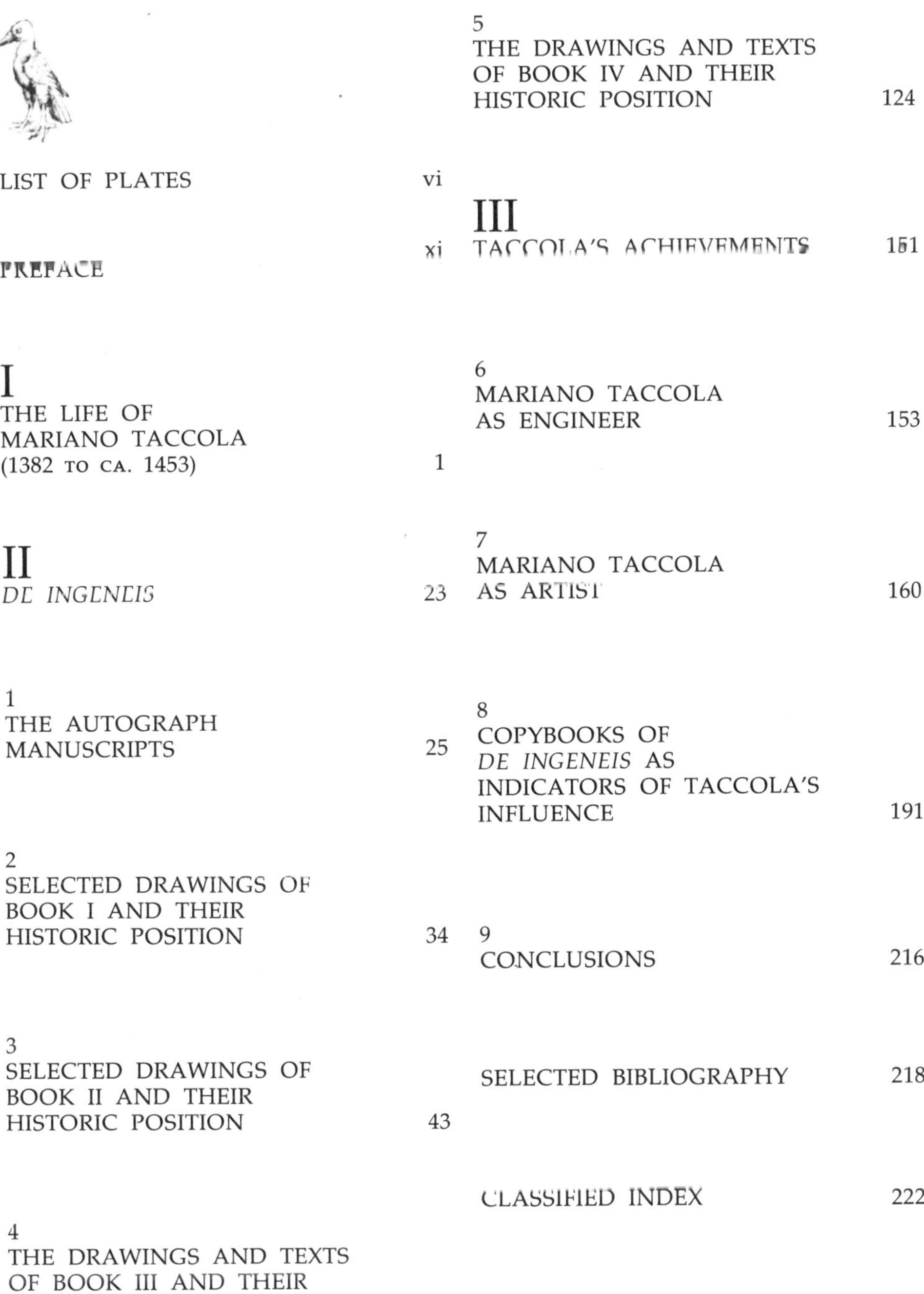

PREFACE

For nearly two hundred years it has
been known that Mariano Taccola was
active in the first half of the Quattrocento, held various offices, and did engineering work that earned him the name
of "the Sienese Archimedes." We know
he was an artist because the *Opera del
Duomo* of Siena paid him for sculptural
work, but the sculptures themselves
remain unidentified. Some of his books
are known, but they have been published
only in small part. What are his real
contributions? Men who studied his
technical drawings and texts have suggested this question, but left it unan-
swered.*

The present authors undertook a review
of primary and secondary sources concerning Taccola, which has led to this
publication. We completed it in this form
in 1966–1967. Scaglia had previously
explored the Taccola manuscripts from a
viewpoint of the history of art. Prager
had analyzed them from a viewpoint of
the history of technology. We transcribed
and translated his texts and considered
the drawings. So far as possible we

* Among these men are M. Berthelot, "Les
manuscrits de Léonard de Vinci et les machines de guerre," *Journal des Savants*, 1902,
pp. 116–120; P. Fontana, "I codici di Francesco di Giorgio Martini e di Mariano di Jacomo
detto il Taccola," *XIVe congrès international
d'histoire de l'art. Résumés des communications
présentées*, Brussels, 1936, p. 102; L. Thorndike, "Marianus Jacobus Taccola," *Archives
internationales d'histoire des sciences*, VIII, 1955,
pp. 7 ff.

traced the historic background of the
man and his works, the biographic facts,
and the vicissitudes of documents and
manuscripts.

Our study revealed the existence of an
unsuspected, early treatise, dated 1433;
we have identified it as *De ingeneis* [On
Engines] by Mariano Taccola. We found
its autograph in a part of a well-known
manuscript in Munich, Codex Latinus
197, and in a little-known but better-
preserved codex in Florence, Codex
Palatinus 766. We also found copywork,
derived from this autograph treatise,
which has received little review thus
far.

The newly identified treatise turns out
to be rather different from the other
parts of the Munich codex, Lat. 197, and
also rather different from Taccola's treatise, *De machinis*, dated 1449, which has
been known in two copies. Incidentally,
we identified the autograph of this treatise and rediscovered a third copy of it.

In view of the newly established facts,
former views about Mariano are in need
of reconsideration. He has been described
as a designer of military devices, but his
manuscripts show that he was at least
equally engrossed in other areas of technology, including power systems for
mills and devices for the building of
foundations. In addition, figures and
landscapes in his autographs show him
as an artist with fine sensibilities for
drawing, not only as an engineer. This
graphic work adds a new dimension to
Taccola's image and a new series of
facts to the history of Quattrocento art.
New problems also arise from Taccola's
recorded works, for example whether
his machines will explain some of the

mechanical work of his predecessors and followers and whether his drawings will allow sharper dating of other artistic work or identification of Taccola's own forgotten sculptures for the Duomo choir stalls.

In arranging for this publication, we received valued assistance and encouragement from many persons. Dr. Eugenia Levi, head of the manuscript division in Florence, facilitated our studies of the Palatine manuscripts in the Biblioteca Nazionale, both before and after the disastrous flood of 1966. We are indebted to Professor Emanuele Casamassima for helping to obtain permission for facsimile publication of Palat. 766, granted by the Ministero della Pubblica Istruzione. In Munich, Drs. Dressler, Gichtel, Montag, and Striedl of the Bayerische Staatsbibliothek were kind enough to provide us with the permit to use material from Lat. 197, and they greatly facilitated our research relating to this manuscript and the later treatise, which is now in Munich and is designated Lat. 28800. Professor Lynn White, Jr., of Los Angeles assisted us most liberally by providing information on critical bibliographic facts and by loaning us for generously extended periods of time his microfilm of Lat. 197. In tracing other parts of the Taccola record, we received valued help from Dr. Wilhelm Mrazek of the Museum für angewandte Kunst in Vienna, Dr. Gino Garosi, Director of the Biblioteca Comunale in Siena, Dr. Ubaldo Morandi of the Archivio di Stato in Siena, and Professor Alessandro Parronchi of the Università di Urbino. Dr. Luigi Michelini Tocci in the Biblioteca Apostolica Vaticana was kind enough to show us photographic enlargements as well as the minute original of the copybook, Urb. Lat. 1757, and to explain his views in advance of his publication of this codex. We are also indebted to Dr. A. G. Keller of the University of Leicester and to Dr. Henry Millon of the Massachusetts Institute of Technology, who have read our manuscript and given it critical review and suggestions. Dr. G. B. Pineider in Florence provided reproductions of Palat. 766, using great skill in recording the drawing, which is hardly visible, on its cover. We also benefited from the kind assistance of librarians and custodians of manuscripts in the New York Public Library, the British Museum, the library in Göttingen, the Bibliothèque Nationale, the Biblioteca Marciana, the Galleria degli Uffizi, and other archives in various parts of the world.

Frank D. Prager
Gustina Scaglia

THE LIFE OF
MARIANO TACCOLA
(1382 TO CA. 1453)

I

Mariano was christened in Siena on 4 February 1382. The baptismal register shows him as Mariano Daniello, his father as the wine dealer Jacopo, son of Vanni called Taccola, and his mother as Madonna Nofria.[1] There was a much older brother Giovanni, who married in 1378, and a sister, called Francesca, who married a silk trader of Pisa when Mariano was eleven. Wine dealers were among the *popolo minuto,* while silk traders belonged to the *popolo grasso.*

The word *taccola* is the name of a small, black, crowlike bird. Mariano inherited it as a nickname or early type of family name from his father, grandfather, and perhaps earlier forefathers. He shows a *taccola* bird as one of the vignettes in his book (Pl. 28 at 30v). Mariano never used his second name Daniello, and the literature has consistently disregarded it. Some modern authors call him Marianus Jacobus, or Jacobus Marianus, or even James. The form Jacobus, instead of Jacobi, arises from misinterpretation of his signature and disregard of the baptismal record.[2] Various examples of his signature can be found in *De ingeneis,* for instance at III 29v and 47v.[3] His correct name is Mariano di Jacopo, detto Taccola.

Siena is built on a hill, some hundred feet above neighboring valleys and a thousand feet above sea level. Inside the city, winding streets go toward and around the central, low-lying, shell-shaped Campo Fori and its Gothic Palazzo Pubblico. On the flanks of the hill are many gardens and vineyards. Farther down, where grains now grow, there were swamps in Mariano's time. Their clearance was one of the concerns of Siena whenever there was peace within her walls and on the borders of her republic.[4] The city itself is high and dry but is surrounded by higher hills, where natural wells could be found. It was another concern of the Sienese to bring water from these wells into their city and to protect the supply from enemies. This they did by underground aqueducts, which they designated with a Greek (or Etruscan) word, *butini* (*bothynoi*), a device of great antiquity. Fountains in the city were fed by these ducts, and water mains crisscrossed the underground between these fountains. They supplied water in amounts unheard of in other medieval towns, except Rome.[5] Siena was able by means of this water supply system to operate the large *Spedale della Scala,* one of the greatest organizations of the time.[6] Only envious

---

[1] Milanesi, *Documenti,* II, pp. 284–286; P. Bacci, *Francesco da Valdambrino,* Siena, 1936, pp. 367 f.

[2] He usually signs *ex manu Ser Mariani Jacobi decti Taccola.* The signature alone could mean either (1) "from the hand of *Ser* Marianus Jacobus, called Taccola," or (2) "from the hand of *Ser* Marianus—son of Jacobus—called Taccola," or (3) "from the hand of *Ser* Marianus, son of Jacobus called Taccola." The record shows that the first of these interpretations is wrong. Obviously there is no basis whatsoever for "Jacobus Mariani."

[3] When citing the various books of *De ingeneis,* we use Roman numbers. The Arabic numbers, thereafter, stand for the folios. Further details are noted on p. 27.

[4] There is no adequate history of Siena. For general orientation, see V. Buonsignori, *Storia della repubblica di Siena,* Siena, 1856, G. Milanesi, *Sulla storia civile ed artistica senese,* Siena, 1862.

[5] Bargagli-Petrucci, *Le fonti,* I, pp. 63–76.

[6] A. Liberati, "Lo Spedale della Scala," *Bulletino senese di storia patria,* n.s., X, 1939, p. 160 f (hereinafter, *Bulletino senese*).

neighbors, including Dante of Florence, were moved to ridicule Siena's frequent construction and repair of water mains.

Another proud institution of Mariano's town is her *Opera del Duomo*. The cathedral accommodates many thousands, and the citizens had once planned to make it several times larger. They also insisted on making it a shrine of unequaled splendor.[7] Meanwhile the governing plutocrats had proud, tall palaces built for themselves. In the small harbors of the Republic of Siena, ambitious plans were underway during Taccola's time to compete with the famous ports of Genoa, Pisa, Naples, and Venice. The plans became an important challenge for Taccola and his followers, as they gave rise to technical problems.

Since the Republic was rich, her neighbors were envious. There was much friction with Rome in the south and Florence in the north, and cruel wars were fought. The Empire claimed Siena as one of its "free" cities[8] but was unable to protect her. Compared with the scale of earlier and later wars, the conflicts of the time were only local affairs, but to the participants they were large. Colossal figures arose and fell in those wars. Therefore, in Taccola's world the engines of war were as important as they have been at most times—even if they were puny compared with either ancient or modern devices. The retrogression and stagnation of technology that had come with the Dark Ages was drawing to an end.

Nothing is, unfortunately, recorded of Mariano's education or apprenticeship,

and we can only visualize a few isolated scenes on the basis of contemporary writings. The atmosphere must have been depressing. The city was depopulated by epidemics, famines, and foolish acts of government. The years 1382 to 1385, immediately following Mariano's birth, saw a major breakdown of local rule. The so-called Reformers, who had governed Siena some twelve to fifteen years, lost their power. With them the new rulers exiled four thousand artisans, and this was one-sixth of the population.[9] The effect was that warlike adventurers from north and south, mainly Milan and Naples, became very bold in making inroads, both political and military. In 1399, when Mariano was seventeen, the Duke of Milan became the owner of Siena. His rule lasted five years. Its breakdown was followed by economic recession, which continued during most of Mariano's adult life. It is, however, not clear to what extent these conditions darkened his youth or his later years. An account of the time when he was still a child, quoted in Romagnoli's Sienese chronicle but lost since then, shows that the family lived near Porta Ovile, one of the city gates.[10] We may assume that the

[9] G. Pardi, "La popolazione di Siena e del senese attraverso i secoli," *Bulletino senese,* XXXII, 1925, pp. 17 ff; also see A. Lisini, "Notizie delle miniere della maremma toscana e leggi per l'estrazione dei metalli nel medioevo," *Bulletino senese,* n.s., VI, 1935, pp. 185–256.

[10] E. Romagnoli, *Bellartisti senesi,* IV, p. 339 (hereinafter, *Bellartisti*): *Il padre del architetto Mariano é nominato nel volume 7 della classe F delle reformaggioni anno 1385 in contrada San Pietro Uvile: Jacomo di Vanni del Taccola die avere sol. 24: allo stratto di Giorgio a fol. 271.* The original *estratti* and *reformaggioni* are lost. Romagnoli (1776–1838) wrote the 12 volumes

[7] V. Lusini, *Il duomo di Siena,* Siena, 1911, *passim.*

[8] Hefele, *Conciliengeschichte,* VII, p. 392.

view of the surrounding vineyards was as pleasant as it still is today. Also, there were short-term causes for cheer. For example, the return of freedom from the Milanese, in 1404, may have gladdened the hearts of the Sienese.

This also was the time of the Great Schism in the Catholic Church (1378–1417). Sometimes the people of Siena witnessed attempts, usually originating in nearby Rome, to end the conflict with the Counter-Pope in Avignon. Such an attempt occurred, for example, in 1407, when Gregory XIII, elected in the previous year, "entered Siena to resolve the Schism" and stayed for some considerable time, "granting many indulgences."[11] It was then, and before this Pope traveled on to Pisa (where he was deposed in 1409), that new work was done in Siena's cathedral and that Taccola, age twenty-six, appeared on the scene. A document of 1408 shows that the *Opera del Duomo* then paid him for sculptural work.[12] A sumptuous cathedral choir had been constructed during Taccola's youth, and its ornamentation was renewed in some way—exact details are unknown, although the years 1407 to 1425 have been cited—perhaps by men in the circle of Domenico dei Cori who made the choir stalls of the Chapel in the Palazzo Pubblico in 1415–1428.[13] It appears that Taccola was among

these artists, as the document states that he was paid for *sedici teste per por al chor del altare maggiore*. The relation of these "sixteen heads" to the original ninety choir stalls of the cathedral and to the thirty-six busts of wood preserved in the relocated parts of this choir has not thus far been established. In fact, no one has studied Taccola's artistic position at all.[14] It is not clear, for example, whether he maintained a sculptor's shop either in or after 1408. Surely, however, he had many contacts with artists.

In 1408, Taccola also established a home. He married Madonna Nanna, the daughter of a leather merchant.[15] No doubt he became, in fact if not by entry in a guild membership list, a master of his art. Of children we hear nothing for more than a dozen years. Nor is artistic work done by Taccola recorded in further public documents of those years. However, his name appears in connection with matters of state transacted by the Sienese. The evidence is furnished by one of his artist friends. Bindino da Travale, a painter and chronicler, reports a strange meeting, witnessed by Mariano Taccola in 1413.[16] It took place in Rad-

---

of his manuscript essays for the Biblioteca Comunale of Siena, where the work received signature L II 14.

[11] The event of 1407 is noted in an inscription quoted by E. M[annini], *Guida storico-artistica del duomo di Siena*, 1908, p. 222.

[12] Milanesi, *Documenti*, II, p. 286.

[13] Carli, *Scultura lignea senese*, p. 27. The years 1407–1425 are specified by Z. Pepi, *Il duomo di Siena*, Siena, 1964, p. 130; also see Milanesi,

Documenti, I, pp. 328–383; V. Lusini, "Dell' arte del legname innanzi al suo statuto del 1426," *Arte antica senese*, Siena, 1904, p. 205 f.

[14] In E. Carli, *La scultura lignea italiana dal XII al XVI secolo*, Milan, 1960, p. 63, Taccola is briefly mentioned in connection with the choir, but the author sees no hope of attributing the pieces preserved. This pessimism ought to give way to industrious work in collating the abundant documentary material with the surviving sculptures. So also Bacci, *Francesco da Valdambrino*, p. 81 n.

[15] Romagnoli, *Bellartisti*, IV, p. 348.

[16] *La cronaca di Bindo da Travale, 1315–1416*, ed. V. Lusini, Siena, 1900, p. 240 ff: *Chome lo enbasciadore del re di Francia inchomenciò la*

icofani, a fortress on a steep rock near the southern border of the Republic of Siena. Bindino entitles his chapter "How the Ambassador of the King of France began to speak before Pope Giovanni, speaking dishonest words to him." Bindino then reports these words: "The King of France finds that it is not proper or honest for the Pope to make war and bloodshed between Christians. There is too much banditry. Pope Giovanni should not stay in bed until the hour of the tierce. Also, popes should not be such miserable liars."

The French ambassador flung this assortment of invective at the Pope in the presence of a Count of Palestrina, ambassador of a King "Unzilau," whose name is also written or transcribed "Vinzilau" or "Anzilau" at numerous points where the chronicle refers to him. Bindino further records that Palestrina told the Pope, "Return to Rome and into your estate. Be the lord of your spiritual realm and let King *Vinzilau* be the temporal ruler."

It is also reported that Palestrina told Taccola about these speeches. In turn he told them to Bindino, from whom we take them as Taccola's first reported statement. No doubt Taccola reported them also to the government of Siena. That government, according to Bindino and other sources, then had ambassadors at various places. It had one group of them travel to a "King of Hungary, the Emperor's son, who was at Udine on his way to Rome to take the imperial crown." Who was this Pope and who were these kings?

Giovanni XXIII, the former *condottiere* Baldassare Cossa, was the successor of Alexander V, the Pope or Counter-Pope elected by the Council of Pisa. He was attacked by both France, whose Pope Benedit XIII resided in Avignon, and Naples, an ally of France. The "King of France" was Charles VII, perhaps best known from Shakespeare's *Henry V*. The man given the strange name Anzilau is clearly recognizable as Ladislao of Anjou-Durazzo, a young soldier-king of Naples, who had started an attempt to become master of Rome, Siena, Italy, and the Empire; he did not live more than a year beyond this time.[17] The "King of Hungary" was Sigismund of Luxemburg (1362–1437), a prince whom Taccola was to meet in later years. Sigismund also was a "King of Rome," that is, Emperor-Elect. He was in need of a Pope who could and would crown him in Rome and thereby perfect his claim to the highest worldly rank then known.[18]

---

proposta a dire dinanzi al papa Giovanni, diciendo a lui disoneste parole. . . . Non è chonvenevole nè onesto dalla parte del re di Francia, che 'l papa facci gittare il sanghue in terra nè che stieno chon tanti inbandigioni; nè che si istia il papa Giovanni infino a terza di chonserva infino a terza nel letto; nè anche ch' e' papi sieno buggiaroni. Mariano di Tacchola da loro seppe il tenore. Lo 'nbasciadore dello re Unzilau fu il chonte de Pellestrina, il dicitore. The chronicler died in 1418 (P. Bacci, *Fonti e commenti per la storia dell' arte senese*, Siena, 1944, p. 218).

[17] The identity of *Vinzilau, Unzilau, Anzilau,* and *Ladislao* (Lancelot) is shown by the *Cronaca*, p. 230 ff. Perhaps the strange name *Anzilau* is derived from *L'Ancelao* (Lancelao), and the other names are due to errors of the modern editor or of the original scribe, Giovanni di Bindo, the author's son.

[18] For this and other details of German-Italian history in Mariano's time, we generally rely on the standard works listed in the Selected Bibliography, cited here briefly as *Reichstagsakten* (major state papers) and *Regesta imperii* (minor

This is what the chronicle meant when it stated that this King of Hungary was "on his way to Rome."

At the time of the meeting in Radicofani, one could say that Rome was "on its way to Sigismund." This German prince appeared in Northern Italy as a result of a war that he was waging, from time to time, against Venice; he would have been willing but not able to travel farther south. However, Pope Giovanni had been driven out of Rome by Ladislao and had fled to Siena.[19] The city, which also maintained relationships with other powers, preferred to receive him in Radicofani.[20] From there, Pope Giovanni soon traveled farther north to meet the Emperor-Elect in Lodi near Milan. The two called the Church Council of Constance. This Council then convened in the next year and rather promptly deposed Giovanni, also removed Benedict of Avignon, persuaded another Counter-Pope Gregory XII to resign, and elected Martin V. The Council and Sigismund are also remembered for the treacherous murder of John Hus, an act that precipitated the Hussite wars and a series of imperial disasters in these wars.

In Taccola's smaller world we hear nothing more of attendance at meetings of kings or popes. In 1416, Siena acquired the city and territory of Chiusi, not far from Radicofani, where bridge and irrigation works were going on and where the boundary of the Papal State needed to be watched. Mariano, along with others, may have learned of construction works in the field of civil engineering about this time. However, the known documents deal with other matters. In 1417, and again in 1420, Mariano was nominated as applicant for entry into the Sienese "Guild of Judges and Notaries." This presupposed that he had been apprenticed to a notary or had studied notarial law at the *Studio*. Either study or apprenticeship generally lasted some six or seven years, according to Sienese practice; but so far as the rules of the various guilds are known,[21] they were not rigid in this respect, and it is conceivable that Taccola, seeking membership in a second profession, could take the examination required by statutes after a relatively short period of preparation. In any event he took the examination of the notaries' guild and twice presented his application for membership. Each time he was admitted but then promptly removed from the membership, since he failed during the ensuing year to appear for final approval of his entry, as was also prescribed.[22] The reasons for his failure to appear are not reported. Later documents, both private and official, call him *Ser Mari-*

state papers), also Aschbach, *Sigismund*, I, pp. 1–350. For general orientation, see J. Bryce, *The Holy Roman Empire*, London, 1864; New York, 1961, pp. 229–251.

[19] See for example, Aschbach, *Sigismund*, I, pp. 371 ff; Hefele, *Conciliengeschichte*, VI, *passim*.

[20] Giuseppe de Novaes, *Elementi della storia de' sommi pontefici*, Rome, 1775–1792, V, p. 48. This detail is not clearly stated in the standard works. In Aschbach, *Sigismund*, the Council of Constance is described in II, *passim*.

[21] F. L. Polidori and L. Banchi, *Statuti senesi*, Bologna, 1863–1877, I, II, *passim*.

[22] Milanesi, *Documenti*, II, p. 286. Regarding the statutes, Zdekauer, *Constituto*, p. 249, and G. Prunai, "I notai senesi del XII e XIV secolo," *Bulletino senese*, LX, ser. 3, XII, 1953, pp. 78–109, especially pp. 86–89.

ano,[23] and the word *Ser* was a notary's title. However, this title was also used by holders of office in church or in charity, and Mariano held this latter kind of office. Perhaps he applied legal skill to become what may be called a notary-in-fact, or perhaps his study already was interrupted by deliberations in the field of technology. Also, the times continued to be agitated.

The year 1423 brought the event known as the Council of Siena, a further attempt to reform the church. The new Pope Martin V, elected at the Council of Constance, was invited to this new Council, where the reforms discussed earlier were to be given further and perhaps more positive consideration than they had received. He refused to come; he found too much independence in the Church Councils then convening. Sigismund did not come to Siena either, as he was short of money and unsuccessful in war, but he maintained contact with the Council from abroad.[24] He promoted his imperial claims vigorously, as his stepbrother, ward, and rival Wenceslas had died. Neither Pope nor reformers nor Sigismund made any progress at the Council, but still another Council was called to open in the city of Basel a few years later.

Directly after this Council of Siena, and possibly due to local events no longer recognizable from the record but somehow connected with this gathering, Taccola began to act as a public official. In 1424, he became secretary, *camerarius,* of an institution called *Casa di Misericordia e Sapientia.*[25] Before his time this Casa had been founded as a hospital, then only called Misericordia.[26] During Mariano's youth it gradually developed into a home for poor scholars, later only called Sapientia, although it still did some hospital work. The attendants belonged to the Tertiary Order of the Humiliates and were housed in S. Maria della Misericordia, also called San Tommaso or connected with a church of that name.[27] This complex of San Tommaso, Misericordia, and Sapientia was part of the larger and still more heterogeneous organization known as the *Studio,* which served as the University of Siena. It was governed as well as financed, by clerical and civic powers, including the Bishop of Siena, the *Spedale della Scala,* the *Opera del Duomo,* and the municipal administration of customs or *gabelle.*[28] As secretary of the Sapientia, Mariano Taccola must have been in contact with influential men of these powerful institutions. On their behalf he probably performed such du-

[23] Milanesi, *Documenti,* II, p. 165.

[24] Hefele, *Conciliengeschichte,* VII, pp. 353 ff, 389 ff; N. Valois, *Le pape et le concile, 1418–1450,* I, Paris, 1909, pp. 20–75; N. Mengozzi, "Martino V ed il concilio di Siena," *Bullettino senese,* XXV, 1918, pp. 247–314; *Reichstagsakten,* VIII, *passim* and X, pp. 279–313; *Monumenta conciliorum,* p. 83.

[25] Romagnoli, *Bellartisti,* IV, p. 339; Milanesi, *Documenti,* II, p. 286.

[26] Zdekauer, *Studio,* pp. 31 ff; G. A. Pecci, *Relazione delle cose più notabili della città di Siena,* Siena, 1752, p. 136. For general orientation: H. Denifle, *Die Entstehung der Universitäten des Mittelalters bis 1400,* Berlin, 1885, pp. 450 f; G. Prunai, "Lo studio senese dalla 'migratio bolognese' alla fondazione della Domus Sapientiae," *Bulletino senese,* LVII, ser. 3, IX, 1950, pp. 3–54.

[27] Tiraboschi, *Vetera humiliatorum monumenta,* I, pp. 180 f; II, p. 106; III, p. 186 (hereafter, *Monumenta*).

[28] Zdekauer, *Studio,* pp. 81 ff, 119 ff, 161 ff.

ties as levying entrance fees from newly arriving students, keeping account of borrowed books, and giving daily instructions to servants, as was usual in an institution of this type.[29]

In the early nineteenth century, the archives of Siena still preserved documents, now lost, which were known and quoted by Romagnoli. He writes that Mariano's numerous memoranda as secretary of this institution always were "approved in public to the sound of a trumpet, on the steps of the Palazzo del Notaro." These memoranda by Mariano Taccola started in 1424 and covered the next seven years; thereafter Giovanni Patrizi became his successor."[30] We may visualize these men in Humiliate attire, brown tunic and black cloak.[31] Mariano's state of wedlock was compatible with his membership in this semimonastic, charitable order. As artist in the service of the cathedral, attendant at the meeting of Radicofani, and holder of a significant office, Taccola must have attracted local interest even before he became known as being, in addition, a builder and inventor.

In 1426, Mariano and Madonna Nanna had a daughter, whom they called Alba. The child's godfather was Jacopo della Quercia, the famous Sienese sculptor. He happened to be out of town on the day of christening, and his place was taken by Pietro di Maestro Giovanni Fittoso, a physician and banker.[32] Mariano had begun to move among the important people. He may have met Donatello, who then worked in Siena. His recorded relationships to the Notaries' Palace make it most probable that he met Gentile da Fabriano, who then painted a much admired picture for this palace. At the Sapientia, Mariano was in perfect position to meet men of learning, even if specific items of evidence for such meetings are rare. Of the leading humanists, Filelfo then was in Siena and probably at the Sapientia.[33] Panormita, a humanist most active at the Studio and Sapientia, made one of the major scientific rediscoveries of the time when he found, in 1426, the *Libri medicinae* of ancient Celsus among the forgotten books of a Sienese monastery.[34] In the learned circle close to Mariano, the sensation must have been great. There was similar sensation throughout Italy when someone found a good and clear copy of ancient Vitruvius, or of the famous mathematicians, physicists, or astronomers,[35] or a lost work of the famous

---

[29] L. Banchi, *Statuti de la Casa di Santa Maria de la Misericordia di Siena*, Siena, 1886, pp. 29–32; *idem*, *Statuti volgari de lo Spedale di Santa Maria Vergine di Siena*, Siena, 1864, p. 19.

[30] Romagnoli, *Bellartisti*, IV, p. 339. He states that he found the memoranda in volume 6 of the *Revisioni delle ragioni di camerlinghi uffiziali*. The original volumes are now lost, according to information from the Archivio di Stato in Siena.

[31] Tiraboschi, *Monumenta*, I, pp. 227 f.

[32] Bacci, *Francesco da Valdambrino*, pp. 367 f.

[33] Guarino of Verona, writing to Siena, referred to a most famous *physicus* Mariano. See R. Sabbadini, *Le scoperte dei codici greci e latini*, I, Florence, 1905, p. 131 n. It is possible that Guarino referred to Mariano Taccola, but he may have intended Mariano di Ser Jacopo Manni, a physician, who is listed in A. Garosi, *Siena nella storia della medicina*, Florence, 1958, p. 383.

[34] C. Corso, "Il Panormita in Siena," *Bulletino senese*, LX, ser. 3, XII, 1953, pp. 139 ff; Zdekauer, *Studio*, pp. 40, 49; Valois, *Le pape*, I, pp. 74 f, 218 ff; II, 277 ff.

[35] Printed editions appeared later, for example Vitruvius, *De architectura*, Rome, 1486; *Scrip-*

Frontinus.[36] Meanwhile the Sienese undertook engineering works of some magnitude[37] and introduced new industries.[38] There was popular debate about mechanical inventions of the great Florentine, Brunelleschi. To a prepared mind the period was ripe with suggestions for scientific and technical work.

The earliest date in Taccola's extant technical writings is 1427. It occurs in Book I 42r and 61r of Cod. Lat. 197, Taccola's autograph in Munich, which is usually called his notebook. His technical observations and deliberations are likely to be of earlier origin; on one of the pages of his book preceding the first that refers to 1427, he writes about technical knowledge that he had acquired with "long labor" (I 31r). Small wonder that he, the sculptor, secretary, and engineer, did not finish his guild admission proceedings in 1417 and 1420.

In his so-called notebook, which we consider as including part of the autograph of a treatise, Taccola first concerns himself with the building, defense, and operation of harbors (*De ingeneis*, I 1r to 20v). Siena wanted to be counted as a maritime power, and Mariano took an apparent interest in her hopes of grandeur. On his early pages he shows a variety of ferries, ship-loading cranes, harbor chains, and chainlike structures, adapted either to float on the water or to sink to the ground in order to trap enemy ships. He describes and illustrates these constructions in what he calls *Capitula de precipitantibus et torquentibus* [Chapters on [Weapon] Projecting and Tilting Devices], briefly *capitula*. These first sheets of Lat. 197 became, later, the first part of his Book I of *De ingeneis*. Plates 1 to 4 show some typical drawings from these Chapters.

While Mariano wrote these Chapters, he remained in contact with important artistic projects. In 1427, Siena commissioned two local sculptors to make precious, silver angels as a gift for Martin V, the Pope elected by the Council of Constance; one of the artists was Giovanni Turini, a famous silversmith. Taccola, another artist, and two of the highest officials of Siena were appointed "to pursue and expedite the business of this gift."[39] They may have gone to Rome as bearers of the gift, but no such trip is explicitly recorded.

One of Mariano's dated notes of 1427 (Book I 61r) relates to bridge work in Rome and harbor work in Genoa, *in Tibere, in porto Janue*.[40] Did Taccola do such work, or did he inspect it? He

---

tores astronomici veteres, Venice, 1499; and *Veteres mathematici* (meaning military engineers), Paris, 1693.

[36] Sabbadini, *Scoperte*, I, p. 85.

[37] L. Banchi, "I porti della maremma senese durante la repubblica," *Archivio storico italiano*, ser. 3, XII, 1871, pp. 102f; C. Cecchini, "Il piu antico documento senese sul porto di Livorno," *Bulletino senese*, n.s., IX, 1938, pp. 81–85.

[38] L. Banchi, *L'arte della seta in Siena*, Siena, 1881, pp. 135 ff; S. Borghesi and L. Banchi, *Nuovi documenti per la storia dell' arte senese*, Siena, 1898, pp. 120 ff, 146 f, 152.

[39] *Circa locationem dicti doni . . . locare et expedire* (Milanesi, *Documenti,* II, p. 137). There is evidence of friendly relations between Giovanni Turini, Francesco da Valdambrino, Domenico de' Cori, and Lorenzo Ghiberti (*ibid.*, p. 120).

[40] Compare: C. von Fabriczy, *Filippo Brunelleschi*, Stuttgart, 1893, pp. 337–382; G. Gaye, *Carteggio inedito d'artisti dei secoli XIV, XV, XVI*, Florence, 1839, I, pp. 555–570. A few pertinent facts are also stated by G. Guarnieri, *Da Porto Pisano a Livorno città*, Pisa, 1967, pp. 142–144.

states only that he built and tested engines, showing that the work "can" be done. Other notes, without date, mention hydraulic constructions in Toledo, in Venice, and elsewhere.[41] A copybook (Add. 34113, 32r) identifies one of the works as located in Livorno. Mariano may have met Giovanni da Siena, then known as *summus architectus*, or his successor Sano di Matteo da Siena, men who concerned themselves with river works, a type of work that continued to be done in various parts of the Republic of Siena, both east and west of the city.[42] As officer of a learned institution, Mariano was likely to have some contact with men who planned such work, even if he did not spend much of his time in the field. Clear it is that he took a strong interest in the work and that he learned and recorded rules and experiences that had previously been transmitted orally and by practitioners.

Probably the most remarkable of his workman informants, one whose experiences were most understandable to Taccola, was Brunelleschi. His meeting with this great man stands out in greater sharpness than other events of his life. Taccola reports it as follows (Lat. 197, S 107v, 108v):

Pippo Brunelleschi of the great and mighty city of Florence, a singularly honored man famous in several arts, gifted by God especially in architecture, a most learned inventor of devices in mechanics, was kind enough to speak to me in Siena. He spoke these words: Do not share your inventions with many persons; share them only with men who understand and love science. If you disclose too much about your inventions and achievements, you give away the fruit of your genius. When listening to the inventor, many people belittle and deny his achievements, so that no one in honorable places will ever again listen to him. Then, after some months or a year, these persons use the inventor's words in speech or writing or design. They boldly call themselves inventors of things that they first condemned, and they take for themselves the glory that rightly belongs to the inventor. Another type of character is the great big ingenious fellow who first hears about an innovation or invention, then declares it very surprising and calls it ridiculous. He even tells the inventor: Go away, do me the favor and keep such things to yourself, otherwise people will call you a beast. God's gifts to us must not be divulged to envious and ignorant people who ridicule them. We must act as men of wisdom, strength and ingenuity. We must not show the crowd our secrets about waters flowing in ocean and river, or the devices that work on these waters. Let a council convene, with an assembly of experts and masters in mechanical art, to discuss plans and construction of the work. Everyone—the educated ones and the morons—wishes to hear the proposal. Intelligent men understand it; they understand at least something, partly or fully. Morons and inexperienced men understand nothing, not even when things are explained to them. Ignorance promptly moves them to anger; they remain ignorant, although they want to show themselves intelligent, which they are not. These men persuade morons to think as they, feebleminded, do and to scorn intelligent men. Block-

[41] The *Tevare* or *Tevar* is mentioned at I 27r and S 114r and the *Po* at II 81v; also *Toleto* or *Toleti* at I 31v, S 107v; *Venez[i]a* at II 87r and *Segrabia* (Zagreb?) at S 105r.
[42] Milanesi, *Documenti*, I, p. 83; II, pp. 104, 127. About the restoration of the Maremma, ibid., II, p. 358; Zdekauer, *Studio*, p. 44; D. Bizarri, "Tentativi di bonifiche nel contado senese nei secoli XIII–XIV," *Bulletino senese*, XXIV, 1917, pp. 131–160; D. Marrara, *Storia istituzionale della maremma senese*, Milan, 1961.

heads and morons can do much harm in questions about aqueducts, the means for forcing water, their subterranean and terrestrial ascending and descending, and about buildings in or over salt or fresh water. We love those men who know about these things. We ought to keep away from those who know nothing. The headstrong charlatan should be sent to war. A council should be formed only of wise men, who bring honor and glory to the republic. Amen.

This remarkable tirade is followed by a statement about riverworks and millworks and technical problems encountered therein. Instructions are given about how to overcome difficulties that arise from the river itself and from the hazards of inadequate planning and accounting. The reported views are interesting for the story of Brunelleschi, as well as that of Taccola, since no other utterance of this type from either man has been recorded. The general spirit of the speech is in agreement with the image of Brunelleschi drawn by Vasari. Obviously the speaker is a man of self-reliant character, secretive about his ideas but of torrential eloquence about his proposals. In his speech, as in many of his works, Brunelleschi draws sharp and clear lines. It is one of Taccola's claims to historic significance that such a man should have honored him with his confidence. That Taccola expanded his early *capitula* into the *De ingeneis* may be due to the encouragement that he derived from the meeting.

It does not follow that Brunelleschi gave good advice. The secrecy of a "council of the wise" could hardly aid in developing a new idea; the much greater secrecy of the inventive mind alone was needed at that stage. Surely no secret council could be of assistance in introducing a new reality. Such was the unfortunate truth at the time, although inventors tended to close their eyes to it and to seek self-defense in secrecy; Brunelleschi had experimented with a privileged or patented invention, had met with embarrassing failure,[43] and sought self-defense. Inventors still meet the same unfortunate truth today under the shortcomings of the patent system. They still often seek the self-defense of secrecy, and it still is self-defeating, except possibly for most limited stretches of time. They must deal with the public, although it generally appears to them as largely composed of fools and thieves.

Mariano Taccola, as a man capable of creative thought but dependent on worldly gain in a society that offered no protection to thinkers, tended to sympathize with Brunelleschi's sentiments. However, his book shows that there was a stronger motive in him and that he was a born teacher. He adopted the policy of secrecy to the extent that he used "veiled speech" at many points and carefully limited his disclosures. Yet he sought readers and could not help but speak to them with clear specifications and drawings. He often endeavored to develop even greater clarity. He followed Brunelleschi's advice in his mind but not in his heart.

The "secrets of the waters," mentioned by Brunelleschi, is probably an allusion to the inventions that Taccola claims for himself in his note of 1427, written

[43] We are referring to the beginning stages of the failures encountered with his *Badalone*. For details: Prager-Scaglia, *Brunelleschi*, Part 6.

in Book I 61r.[44] For three inventions
this note gives little but the results,
although all of them clearly have to do
with "the waters." For a fourth hydraulic
invention the note gives positive information. In addition, Mariano's later
book gives comprehensive views at
least of the third invention (III 33v–34r,
Pl. 34) and of the fourth (III 43r, Pl. 50)
As to all four inventions the note of
1427 states that "tests" were conducted
and completed. Official tests of inventions were usual and were sometimes
followed by the grant of public support.[45]
It seems probable that Mariano sought
this kind of support and that Brunelleschi's speech was intended to recommend caution in the more or less public proceedings.

It may have been in response to such
counsel, and in pursuit of secrecy, that
Taccola devised small operating models,
such as the mercury-driven replica of a
mill power system (Book I 74r). However, some of the test equipment was
likely to be large and incapable of secret
handling, as it included for instance
caissons large enough to contain foundations for a building. Taccola was sometimes surrounded by helpers, both
wanted and otherwise, in the construction and floating of such devices. Surely,
however, "he did not waste words."

In addition to his early *capitula* on

harbors and naval war, Mariano had set
down a group of paragraphs on five
pages (II 77r–79r), substantially copied
from an earlier *liber ignium*. This Book
of Fire, which he correctly credited to
Marcus Graecus, a Byzantine writer of
unknown date, dealt with incendiaries,
gunpowder, and illumination. It is a
chemical complement to the mechanics
of the Chapters. Then at some time,
perhaps 1427, the Chapters and the Book
of Fire became opening parts of what
Taccola called, respectively, *Liber primus
leonis* and *Liber secundus draconis*, titles
that he cited in subsequent books. Somewhat later he started his Third and
Fourth Books (Palat. 766) and called
them *Liber tertius de ingeneis ac edifitiis
non usitatis* [On Engines and Devices not
(yet) Used] and *Quartus liber de edifitiis
cotidianis* [On Common Devices]. Near
the end of the latter he stated that he
wrote these books at the Sapientia,
finishing them on 13 January 1433 (IV
73v). A less detailed and less well preserved entry at the front of Book III
seems to refer to a starting date in 1431
and, more clearly, to a historic event
that he calls the arrival of King Sigismund. Apparently he means the king's
arrival in Italy, which took place in late
fall of 1431.[46] That year also brought the
election of a new Pope, Eugene IV, the
beginning of a new Church Council, in
Basel, and the ultimate steps of Sigismund in negotiating with Rome and the
Councils. He had been "on his way to
Rome" as early as 1413, by now he was
definitely expected.

The manuscript of Taccola's *Liber tertius*

[44] Brunelleschi's own designs of the same type
are mentioned by A. Manetti, *Vita di Filippo
di Ser Brunellesco*, ed. Toesca, Florence, 1927,
p. 74; Fabriczy, *Brunellecchi*, p. 612; and P.
Sanpaolesi, *Brunelleschi*, Milan, 1962, pp. 99 f.
Details are in need of study.
[45] Prager (1950), pp. 472–476, 492 f, 519 f; *idem*,
"The Examination of Inventions from the
Middle Ages to 1836," *Journal, Patent Office
Society*, XLVI, 1964, pp. 268 f.

[46] *Regesta imperii*, IX, 2, p. 202; Aschbach,
Sigismund, III, pp. 400 ff.

and *Liber quartus* is dedicated to Sigismund. Mariano may have intended even his early Chapters for such a visitor and may have planned during the 1420s to develop some formal offering or some books of more finished character than his Books I and II, which had gradually become somewhat disorganized. Then the year 1430 brought an outbreak of the plague, during which time the Studio and Sapientia were moved temporarily from Siena to neighboring Lucignano, in the countryside, as was usual at such times.[47] In the enforced leisure of this exile, Mariano may have begun to plan Book III, but he can hardly have begun to execute it before normal conditions were restored, enabling him to make his frequent references to his own earlier books and, no doubt, to works of others.

Taccola's dedication and other statements in Books III and IV reflect an attitude toward Sigismund that amounts to total submission. Clearly he identifies himself with those citizens of Siena who greeted Sigismund as the restorer of the church and the leader of the world. In a similar way Dante had idolized Sigismund's great-grandfather Henry VII, and more recently Petrarca had toyed with the idea of so idolizing Charles IV, the father of Sigismund.[48] Taccola, obviously influenced by such traditions, expressed the hope that Siena would prosper *sub imperii libertate* (III 33v). The Emperors in turn sought the goodwill of friendly cities and citizens. Henry, in Dante's time, had employed Guido da Vigevano, a physician and engineer who later found similar employment at the French court. Charles, in Petrarca's time, had repeatedly visited Siena and her *Studio.* When Sigismund himself was a young man, he had the engineer Kyeser with him at war,[49] and he knew enough of Siena to make very special inquiry about her hospitals when he received her ambassadors about the time of the meeting in Radicofani.[50] For Taccola, the secretary of a charity that also did hospital work, it was almost obvious to seek the acquaintance of this king. Taccola's dedication (III 69r) includes a request for employment in Hungary.

Sigismund, "on his way to Rome," now arrived in Milan. During the spring of 1432, he visited Lucca, a city allied with Siena, which was being attacked by Florence, an ally of the Pope.[51] About the beginning of July, Sigismund made a dash for Siena with his retinue of some

[47] Zdekauer, *Studio,* p. 40.

[48] G. Voigt, *Die Wiederbelebung des classischen Altertums,* II, Berlin, 1893, pp. 263–270; E. Werunski, *Geschichte Kaiser Karls IV,* II, 2, Innsbruck, 1886, pp. 553 ff; C. C. Bayley, "Petrarch, Charles IV and the *renovatio imperii*," *Speculum,* XVII, 1942, pp. 323–341.

[49] G. Quarg, ed. of K. Kyeser aus Eichstätt, *Bellifortis,* II, Düsseldorf, 1967, pp. XX ff. This war, during which Sigismund tried his luck against the Turks, lasted from 1392 to 1396. It was followed by a falling out between Sigismund and Kyeser. For full determination of Kyeser's influence on Mariano Taccola it would be necessary to consider the copies of *Bellifortis* at Vienna, Munich, and elsewhere. In spite of preparatory work by H. Schulte, recorded in a manuscript (Vienna, Ser. nov. 4850), difficult questions remain to be solved.

[50] Bindino, *Cronaca,* p. 226; Milanesi, *Documenti,* II, pp. 63 f; also see Voigt, *Wiederbelebung,* II, pp. 273–276.

[51] Valois, *Le pape,* I, p. 210; see also *ibid.,* II, pp. 127–158 for the general constellation: Rome and Florence for the Pope; Milan and Naples for the Council; Venice and Siena uncommitted.

500 riders and 1,000 horses, carrying small bombards and other weapons. He narrowly avoided the Florentines. He reached Siena on 11 July. He then stayed in Siena nine months. The visit coincided with final stages of Sigismund's negotiations with Pope and Council, whereby he established powers for himself and his successors and probably contributed to the survival of the church.[52] In Siena, the imperial visitor needed lodging and board for himself and his retinue, while the city wanted help against Florence and the Pope.[53] Some anecdotes of the visit are preserved in the erotic-turgid novel *De duobus amantibus* by Enea Silvio Piccolomini.[54] The secretary of the Sapientia probably knew of the bedroom stories that are the substance of this novel; the work is dedicated to *Ser* Mariano Sozzini, an acquaintance of our Mariano. Incidentally, the Sapientia provided beds for the visiting court.[55]

Sigismund was in the habit of seeing visitors frequently and declaring them to be his *nobiles familiares* or, in due course, his *comites palatini*, as Enea Silvio and others noted with some sarcasm.[56]

Taccola expressed a desire to become a miniaturist, engineer, and a "familiar" of Sigismund (IV 69r, Pl. 97), and he obtained the latter rank during the king's visit. This is shown, independently of all Italian records, in the *Regesta imperii*, under date of 18 November 1432.[57] The document calls him Marianus Jacobi *de humelis*, one of the Humiliates. Thus it reflects his position in a hospital-connected institution. Taccola was the only one so appointed at the time. Two Sienese had received the rank of *familiaris* a week before, and one received it five weeks later.[58] Taccola states (III 27r) that machines were discussed when he spoke with the king. They could converse without interpreter, as Sigismund spoke Latin and Italian.[59] Taccola not only spoke to the king but also was allowed to make portraits of him (Pls. 20, 22). The stature of Taccola must have risen greatly in the eyes of the Sienese. He was well on his way toward becoming their Archimedes.

On 13 January 1433, Taccola applied the *finit* to his book *De ingeneis*, obviously

---

52 Valois, *Le pape*, I, pp. 211 ff, 242 ff; Aschbach, *Sigismund*, IV, p. 1 ff.
53 *Reichstagsakten*, X, pp. 348 ff, 357 ff; F. Merletta, "Una lettera volgare di Mariano Sozzini," *Bulletino senese*, ser. 3, I, pp. 22–36; Valois, *Le pape*, I, pp. 211, 218 f.
54 Written about 1444; frequently printed, beginning about 1471, Cologne. A good modern edition is by J. J. Dévay, Budapest, 1903.
55 Romagnoli, *Bellartisti*, IV, pp. 144, 206, 245, 334.
56 Enea Silvio, *De duobus amantibus*, ed. Dévay, p. 42, mentions one who receives a *comitatus* as *lenocinii mercedem*. Another contemporary, cited by Voigt, *Wiederbelebung*, II, p. 276, called the Emperor *vir in donandis insignibus facillimus*. For the form of grants of

the rank of *nobilis familiaris*· *Regesta*, IX, 1, p. 20. In the later *regesta*, editor Altmann omits the term *nobilis*.
57 *Regesta imperii*, IX, 2, p. 225, citing a signed imperial document of 18 November 1432, bound in Notarisches Reichs-Register J, fo. 195r.
58 *Regesta imperii*, IX, 2, p. 233. They were Nicolaus Salomon, Dr. jur. Johannes Legants, and Johannes Laurentii.
59 Voigt, *Wiederbelebung*, II, p. 275, reports that Sigismund spoke Latin and Italian among other things. Giovanni Cavalcanti, *Istorie fiorentine*, I, Florence, 1838, p. 485, omits the Latin but adds French, Bohemian, Hungarian, Turkish, and Tartaric. Perhaps he also spoke some German. Enea Silvio (*De duobus amantibus*) adds that his Latin was somewhat elementary.

planning to give the work, or a copy of it, to Sigismund. Three months later the king departed from Siena, having completed his difficult negotiations and having exhausted if not overstayed his welcome at Siena; he had paid for her hospitality by ending the war with Florence and the Pope, an event to which Taccola alludes (IV 73v). The king was on his way to the usual ceremonial entry into Rome and to his coronation as Emperor.[60] In the Eternal City he stayed a few months. He then proceeded back north, along the east coast of Italy. On 21 September 1433, he reached Mantua, his last major place of sojourn in Italy.

In this city he then, two days later, once more met a group of Sienese. They may have accompanied him bodily or, as is more probable, only in spirit and by messages, appearing now to take leave of him. According to the *Regesta*, on 24 September, Marianus Jacobi *"de Thomasinis,"*[61] a term that was equivalent to the former *de Humelis*,[62] and other Sienese men became *comites palatini*. No doubt the title was given in lieu of payment for some services rendered or properties delivered, and although the details again are not recorded, probably in Mariano's case the delivery of a copy of his book, personally or by proxy, was the occasion. However, no such copy is known to exist.

In 1434, Mariano Taccola then became *stimatore*, or estimator, in Siena, while his tenure at the Sapientia came to an end. The reasons for the change as well as the exact duties of the new office (perhaps connected with tax matters as in other cities)[63] are unknown, but Taccola's stature in Siena continued to grow. Beginning as an obscure sculptor and scribe, he became one of the most famous representatives of his home town. Contacts with artists continued. Directly after Sigismund's visit, Taccola was in contact with a painter, Domenico di Bartolo, who is named in a document of the *Opera del Duomo* in connection with an official portrait of Sigismund.[64] In 1435, Domenico and Taccola jointly witnessed a document, needed by still another artist known to Mariano.[65]

A humanist was among those who supplied Taccola with graphic material. This was Ciriaco d'Ancona, collector of inscriptions and guide to Sigismund in Rome. This learned man addressed him *elegantissime Mariane* in a letter, sending him sketches of exotic animals.[66] Meanwhile, in 1441 and 1442, Mariano Taccola produced new sculptures, according to payment records of these years.[67] They mention numbers of gargoyles for the choir of the Duomo and a *testuccia*, perhaps a headpiece or baldachin. He also explained his machines, at least to a select public. According to notes of 1438 (Book I 70r, II 82r), the year after Sigismund's death, he gave such explana-

[60] *Regesta imperii*, IX, 2, pp. 234 ff; *Commissioni di Rinaldo degli Albizzi per il comune di Firenze*, ed. C. Guasti, III, Florence, 1873, p. 515.
[61] *Regesta imperii*, IX, 2, p. 248.
[62] Tiraboschi, *Monumenta*, I, pp. 180 f.
[63] *Storia di Milano*, VI, Milan, 1955, p. 498. There is no reference to *stimatori* in Zdekauer, *Constituto*, nor in G. Prunai, "Notizie sull' ordinamento interno delle arti senesi," *Bulletino senese*, n.s., V, 1934, pp. 365–420.
[64] Milanesi, *Documenti*, II, p. 161.
[65] *Ibid.*, p. 165.
[66] C. Mitchell, "Ex libris Kiriaci Anconitani," *Italia medioevale e umanistica*, V, 1962, p. 285 and Pl. XXII.
[67] Milanesi, *Documenti*, II, p. 286.

tions to the leading, local lawyers Daniele Nicholai Romanellis, Mariano Sozzini, and Pietro de Micheglis. The second note states that De Micheglis, who had become *Comes*, now transmitted part of the information to military men. In 1441 (II 96r), Mariano showed his work to a cleric from Spain, Antonius Catelanus. He may also have seen Alberti, coadmirer of Brunelleschi, who was in Siena in 1443 for seven months serving Eugene IV.[68]

In 1441, Mariano Taccola became *viaio,* superintendent of streets.[69] The appointment may reflect a public recognition of his abilities as civil and hydraulic engineer.[70] By then he had long been entering many new drawings and sketches, and some new descriptions, mainly of *apparatus belli* and of some types of hydraulic construction, in his Books I and II, on blank sheets and margins, and in a Sequel to these books. Taccola's new graphic work culminated in a new treatise, the *De machinis,* planned in ten books and the autograph of which we have identified as Lat. 28800 in Munich. This work was completed in 1449, specifically in August of that year, as stated in the copy in New York (Spencer Ms 136, 102r). The general relationship of drawings and texts is similar to that in *De ingeneis,* and again Taccola often completed drawings before their text or without ever adding the text. In the *De machinis,* he shows greater assurance than before. His preface for this work, no longer found in the autograph but preserved in the copy in New York, reads as follows:

This book is the salvation of soul and body, and is the perdition of soul and body. It is salvation to those who always desire to work for love and charity of holy faith of our Lord Jesus Christ, by word and works, against the infidel and barbarian people, and to fight for the truth of said faith and for all Christianity. Perdition it is to those who do not so desire faithfully, but rather try as bad Christians always to usurp and reduce the regions of kings, princes, dukes, counts, margraves and communes and countries of others that live under law, in liberty. Such Christians injure all Christianity by their wars, and are the cause that the infidels in their vicious faith grow in confidence and increase their domination to the damage and vexation of all Christianity. However I, *Ser* Marianus Taccola, also called Archimedes of great and magnificent Siena, have not designed these engines, machines and weapons that they may operate against Christians, but have invented, composed and designed them that they may go against the infidel and barbarian people. I, the aforesaid *Ser* Marianus, am old and weak and cannot personally work against said infidel and barbarian people. However, so long as I live, I wish and desire—God granting—to exercise my powers. I present myself to the Christians, to serve as I can.

The author dedicated the new work to "the Christians" rather than to a prince. Emperor Sigismund had left Italy in 1433 and had died in 1437. He was followed by his son-in-law Albrecht of Austria, and then by Albrecht's son Frederick III, the first emperors of the Hapsburg house and the last to seek coronation in Rome.[71] Frederick appeared

---

[68] G. Mancini, *Vita di Leon Battista Alberti,* Florence, 1911, p. 255.

[69] Milanesi, *Documenti,* II, p. 285.

[70] Zdekauer, *Constituto,* p. 444 (s.v. *viarius*).

[71] F. Gregorovius, *Geschichte der Stadt Rom im Mittelalter,* Stuttgart, 1858–1872, Book XIII,

in Siena in 1452, near the end of Mariano's life. We do not know whether the old *familiaris* and *comes palatinus* saw him. However, a symbol of the old, medieval institutions remained in Mariano's new work and its copies, in the form of numerous, uniform crowns.

During or after the 1440s, Taccola's work as an official of Siena came to an end. However, he continued to receive an income, and he and his wife continued to live in the house provided by the Sapientia. When he was about seventy years old, he became a novice in a new fraternity or order. In a *denunzia* or inventory of worldly possessions, executed by his hand in 1453, he calls himself *friere* Mariano di Jachomo detto Tacchola, and says that he is one of the *aspectanti*, or aspirants, in the Order of San Jacomo.[72] The declaration throws some further light on the course his life had taken. He lists his real estate as mostly rented out to tenants, indicating that title to it was held by his Order while he managed it. Some property had come to him by his wife's inheritance from her parents. Different parcels of land, *poderi*, located near Siena are listed and briefly characterized in terms of income-producing capacity.[73]

Then the statement continues,
I receive 7 florins and 16 soldi a year as provision from the Commune of Siena, and have been paid well. I recommend myself to your Honors. I am old and weak and have no gainful occupation. During the last twenty years I had two occupations as officer of the Commune of Siena, one as *stimatore* of the Commune and one as *viaio*. May God on high preserve you in his holy peace. I, the aforesaid Friar Mariano, can have no office of the Commune of Siena, as I am a member of the Order of San Jacomo and owe obedience to his Highness, Don Fer[n]ando, Master of San Jacomo, in Portugal and in his realms.[74] All my possessions, both real and personal, are under obligation to him and the Order. I must receive the Brothers of this Order in my house, and aid and defend them, the Order and its liberty. I ask your honors to give consideration to my position as a person devoted to religion. I receive no benefit from my Order, as I am among its aspirants.

A mixture of disarming humility and resourceful independence can be read in these lines. Old Mariano is not yet finished with the *ingenia non usitata* that he can muster. In his earlier years he had become an artist and also a notary without membership in either guild.

---

ch. 2; G. Quirm, *König Friedrich III in Siena (1452)*, Göttingen, 1958.

[72] See the whole *denunzia* quoted in Milanesi, *Documenti*, II, pp. 284 f, and a facsimile of a part of the document in C. Pini and G. Milanesi, *La scrittura di artisti italiani*, Florence, 1876, No. 51. Regarding the history of the Order, we know only P. Antonio L. Ferreiro, *Historia de la Santa A. M. Iglesia de Santiago de Compostella*, Santiago, 1904. Romagnoli, *Bellartisti*, IV, p. 337, calls Mariano Taccola "knight" of the Order. If this were taken literally, it would mean that he lived beyond 1453, passing from the rank of *friere* to that of *Cavaliere*.

[73] The declaration gives them such names as *La Scorta* and *Lo Spedaluccio*.

[74] *So de lordine di Santo Jacomo et so a ubiedienza del serenissimo principe signore signore Don Ferando, Mastro di Sancto Jacomo in Portogallo et ne suoi regni.* The statement is enigmatic, as no ruling prince in Portugal and surrounding countries was called Fernando or Ferrante at the time. Taccola may have intended the headmaster of the Order, Fernan Rodriguez de Betanzos, although the expressions used in the declaration appear exaggerated when applied to such a canon.

He had also become an emperor's friend without loss of republican character and an Archimedes without scientific specialization. Now in his old age he became aspirant in a semimonastic order with headquarters in a foreign country, and in the process he arranged to retain use of his properties but to avoid an owner's tax obligation.

Did he turn away from engineering and acquire still other interests? The evidence is limited, but it happens to include an exercise in humorous, literary writing. On folio 104r of Lat. 197, in the Sequel, he recorded the story of a master-turned-servant and his irrepressible magisterial habit. It is one of the stories about Giacomo Colonna, called Sciarra, Roman tribune in 1327.[75] Taccola wrote it in Italian, extremely rare in his notebook, and he wrote it in handwriting that is identical to his *denunzia* of 1453. The story tells of the way master and loyal servant exchanged their respective roles, so that Sciarra, who was forced to wander in exile, would not be recognized by the many persons who sought his death. For years they faithfully enacted the formalities, until the servant-turned-master, having long noted that Sciarra, serving him at table, always kept the best morsels of food for himself, was moved at one time to say, "Any fool can see that you are the master and I am the servant." This homely story, written in elementary prose, may be fact or fiction or may contain political connotations.[76] It seems unlikely that Taccola composed it, but the handwriting leaves no doubt that he wrote it down and that he did it in his old age.

His handwriting is one of the most definite signs of his personality. Throughout *De ingeneis* and *De machinis* it is a Gothic cursive, notarial hand. The handwriting of the second treatise has more force and impulsiveness, but the letter forms and paragraphs are not noticeably different; his hand develops in character but is not changed in substance. The

[75] "E se navedrebbe ghirbi che voi sete el signore / et io sono el servo. Questo con tale dire / et parlare / adivenne in questa forma et caso. Scarra Colonna esendo in exilio perche era ribello di Roma . . . aveva un suo servo . . . seco el menava / come fusse suo signore. . . . Et quando erano ad mensa Scarra stava ricto in pie / ad mensa / el servo sedeva servito da Scarra C di coltello / con tutti e gesti et debite reverentie. . . . Et questo duro piu anni Insieme andarono per lo mondo / fugendo duno reame ad un altro. . . . Ora diveniva alcuna volta chel detto Scarra / tagliando dinanzi al suo servo / esso poneva e bochoni de carne o pesce ed altre carni dilectevogli e migliori senpre dinanzi a sse / et al servo poneva e bochoni non vantagati / ma piutosto da fugigli / che da mangare El servo bene savedeva / et conosceva e vantagi de ghiotti bochoni / molto meglio che suo signore / ma per riverentia / a tale materia poneva silenzio. Et questi cotagli modi fu patiente longho tenpo / et al fine dilibero / di dire a Scarra quando esso mangavano insieme / et serviva esso fameglio di coltello stando ritto el servo sedeva. disse queste parole . con voce bassa et umile al ponare de bochoni in su tagliera se navedrebbe ghirbi / che voi sete el signore / et io sono el servo."

[76] Sciarra Colonna was a symbol of the imperial power admired by Mariano Taccola. It was Colonna, not a Pope, who bestowed the imperial crown on Louis the Bavarian in 1328. The Count of Palestrina, mentioned in Bindino's *Cronaca*, 1413, belonged to this family (see Valois, *Le pape*, I, pp. 103, 114 f). Members of the Colonna family also represented Sigismund in Rome while he was in Siena (*Regesta imperii*, IX, 2, pp. 219 f). Pope Eugene persecuted the family whenever he was not *in exilio* himself (See Gregorovius, *Geschichte*, Book XIII, 1, n. 62).

writing uses letters of regular proportions in body, in stems, and descending strokes; it has a gliding motion, with letters flowing into one another in short succession. At the same time he uses another form of writing in both treatises, adopting it for such special purposes as some titles, short messages, dedicatory inscriptions, and epigrams.[77] In these cases he always suppresses the cursive ligatures. The writing is more formal and rigid; the body of letters is larger and static, the stems shorter and usually devoid of loops. This is lettering rather than writing, the *libraria* hand rather than the *corsiva*.

There are some 280 folios in the *De ingeneis*, its Sequel, and the *De machinis*, written during a period of approximately thirty years. In comparison to the thousands of lines of cursive writing on these several hundred pages, a few dozens of titles or other statements written in the *libraria* script are a minor percentage of the total. However, the entire declaration of 1453 and the entire Sciarra Colonna story are written in lettering form. Are all earlier samples of this type lost? It does not seem probable when such a large body of text is preserved. Thus it seems we may have here a late development. There are mixed and transitional forms of writing.[78] The explicit dates for the various forms of Taccola's hand are too few to justify detailed or rigid conclusions about the age of undated pages or paragraphs, but the material is sufficiently massive and coherent to make it fairly certain that the handwriting as well as the drawing style underwent gradual change, even in his old age.

The moment of Mariano's death and his burial place are unknown. Clear it is that he made the declaration of 1453 in anticipation of an end that he then expected to come soon. For unknown reasons, Milanesi concluded that Taccola must have died before 1458. Milanesi may have thought of a letter that he reported separately from other Taccola documents.[79] This letter was written by Leonardo Benvoglienti to the *cavaliere* Cristofano Felici, *operaio del Duomo* in Siena. It says, "Greetings from my part to master Donatello, master of the portals. Truly he deserves great honor. *Misser* Mariano would have agreed with me on this. Donatello conducted him [Mariano] away from Padua, as long as four years ago; Mariano had a great desire to be in Siena in order not to die near those frogs of Padua, as he almost did. . . . Rome, 14 April 1458. As always yours, Leonardo, ambassador of Siena."

If this was Mariano Taccola, it was his only recorded trip out of Siena, unless he went to Mantua in 1433. There is reference to a Friar Mariano of Siena, who returned from Padua or Venice with precious vases for the Pope in October 1458.[80] Was this the same trav-

---

[77] His early hand may be seen in Plates 1–8, and 21. Samples of the lettering form in titles and inscriptions occur in Pls. 34 (left), and 71 (54v), and, more extensively, in Pls. 93 and 103.

[78] See transitional handwriting in Pl. 14, and on pages of the Sequel (Prager, "A Manuscript of Taccola . . . ," Figs. 2-b, 3-b, 4-b).

[79] Milanesi, *Documenti,* II, p. 299.

[80] E. Müntz, "Les arts à la cour des papes," *Bibliothèque des écoles françaises d'Athènes et de Rome,* II, fasc. 3–4, p. 328. The Pope then reigning was Pius II, the former Enea Silvio Piccolomini of Siena.

eler? The name Mariano is too common
in Italy to prove it.

We have no record of Mariano's appearance. He may have resembled one
of his surveyors, whose picture he drew
on the first page of the last book in his
De ingeneis (IV 58r). His writings, as
well as the picture, project the image
of a man endowed with lively interest
and sharp observation. Apparently he
was a recorder of facts rather than a
man of action or a philosopher. He was
mainly an artist and a teacher.

II

1 The following methods of reference, transcription, and translation are used in this publication.

Reference to manuscript pages is given by number of Book and folio side: I 1r, II 96v, IV 76r, S 97r, SF 29v. The captions to facsimile pages, on the plates of this book, give the full citation of each folio page. The Books, numbered and named by Taccola are identified as follows:

Book I: Lat. 197, fols. 1–21; 30–75
Book II: Lat. 197, fols. 76–96
Book III: Palat. 766, fols. 27–57
Book IV: Palat. 766, fols. 58–76
S denotes a Sequel to Books I to IV: Lat. 197, fols. 97–137
SF denotes a Separate Fascicule now contained in Book I: Lat. 197, fols. 22–29

Books I and II contain the following:
Primary Drawings and Texts, centered on the page;
Sketches and Notes, inserted between and around the Primary Drawings and Texts;[1]

Addenda—a few drawings, centered on the page but showing stylistic development over Primary Drawings.[2]

Taccola's folio numbers, some of which are not visible, are used only when specially needed in addition to modern ones and are shown in parentheses, for example, I 3v (1v) or II 76r (73r).

Latin spelling of medievalisms, Italianisms, and misspellings (e.g., *ttrhare, docieri, comburare*) are unchanged except for Taccola's writing of two-letter forms, which causes us to print *Jacobi* instead of *Iacobi* and *universalis* in place of *vniuersalis*.

Latin abbreviations are resolved. Transcriptions of Taccola's Latin text are italicized. Abbreviations therein are set in Roman type. Examples; *in*, *per*, *dominum*.

Latin punctuation is unchanged.

Use of brackets and parentheses. In the transcription, brackets are placed around Mariano's omissions in order to facilitate the reading, for example, se[m]per. In the translation, brackets indicate editorial additions that make the apparent meaning more explicit. The use of parentheses in the transcription indi-

---

* Since the time we wrote this book, two publications appeared, which are pertinent to the present chapter: P. L. Rose, "The Taccola Manuscripts," *Physis*, X, 1968, pp. 337–346 (confirming some of our conclusions), and J. B. Beck, *Liber tertius de ingeneis ac edifitiis non usitatis di Mariano di Jacopo detto il Taccola*, Milan, 1969 (facsimile of Palat. 766 with introduction).

[1] Berthelot (1891), p. 475, writes of *figures fondamentales* and of *esquisses*. There are about 500 Sketches, if groups of tools and the like are taken as a single Sketch, otherwise several hundred more. About one-half of them are on fols. 50 to 90 of Lat. 197.
[2] Among others, see I 68r, II 90v, 91r, 96v discussed in the chapter "Mariano Taccola as Artist."

cates superfluous letters and words written by Mariano, for example, *serretur porticula(m).*

The Sienese calendar is modernized in the translation.[3] Copies of *De ingeneis* drawings in later manuscripts are listed in footnotes to the corresponding folios of *De ingeneis.*

Lat. 197 in Bayerische Staatsbibliothek, Munich, contains Books I and II, the Sequel, and the Separate Fascicule. The set totals 138 folios, almost entirely written in Latin, on paper measuring 30 × 22 cm. Preceding this set, the codex also contains the so-called Hussite Anonymus, dealing with machines similar to some of Taccola's but illustrating them in very different form and describing them in German on 48 folios of somewhat different format.[4] The codex has wooden covers lined with impressed off-white leather.[5] The paper used by

Mariano has ten different watermarks, including two for Book I, another for the Separate Fascicule, two more for Book II, and five more for the Sequel.[6] There are several systems of pagination in Taccola's part, at least two of which are of modern origin according to their handwriting, while others are of Quattrocento character, cited by Taccola, and no doubt inserted by him.[7] His citations or cross-references into Lat. 197 are of three types: in the early part of Book I, he makes reference from one chapter to another—as more fully described hereinafter—by reference to its title or initial phrase; in Book II, he occasionally cites a Primary Drawing by folio number; and in the Sequel he repeatedly cites Sketches by folio number. These citations or cross-references are rather numerous.[8] Another

[3] F. K. Ginzel, *Handbuch der mathematischen und technischen Chronologie. DasZeitrechnungswesen der Völker,* III, Leipzig, 1914, pp. 161, shows that Sienese dates compare with modern ones as follows:
Sienese 1 Jan. 1381 24 Mar. 1381 25 Mar. 1382
Modern 1 Jan. 1382 24 Mar. 1382 25 Mar. 1382
Also see Y. Renouard, *Les hommes d'affaires italiens du moyen âge,* Paris, 1949, pp. 190 ff.
[4] The Anonymus uses paper measuring 32 × 22 cm. The watermarks are of the *balance en cercle* type (C. M. Briquet, *Les filigranes,* Geneva 1907, nos. 2445–2608, especially near 2500) and seem to be of the middle or later Quattrocento, more recent than is usually assumed. A total of about one-half of the contents has been published by Berthelot (1891), Figs. 1–25, by Beck, *Beiträge,* Figs. 307–333, and Gille, *Les ingénieurs,* pp. 61–68.
[5] Similar in appearance to the covers described by H. Striedl, "Die Bücherei des Orientalisten Johann Albrecht Widmanstetter," *Serta Monacensia,* Leyden, 1952, pp. 236 ff.

[6] In Book I, *Demicerf* (Briquet, *Les filigranes,* no. 3282, date ca. 1427); *Monts en cercle* (Briquet's no. 11853, ca. 1441). In the Separate Fascicule, the watermark *Étoile de 5 points,* in a circle. Book II, *Montagnes* (Briquet's no. 11684, ca. 1400) and *Chalumeaux* (no. 3343, ca. 1424–1442). In the Sequel the watermarks are *Arc, Arbalète,* and three unknown designs, one of which recurs in the paper of Add. Ms 34113, a copybook in London. One of the unknown designs and the *étoile de 5 points* recur in Lat. 28800 in Munich, Mariano's autograph *De machinis.*
[7] There is irregularity of sequence in the first few folios and at various places throughout the codex; therefore we generally use modern pagination. Mariano paginated some folios of the Sequel with numerals 228 to 236 and subsequently changed them to read 107 to 116, these numerals being placed in the upper center of the folios. A group of these folios is reproduced by Prager (1968), Figs. 1–13.
[8] He cites,
I 15v (16v) at III 47v
  18v (20v) " IV 62v
  18v (20v) " II 92bis r
  21r (23r) " S 126r
  21v (23v) " S 126r
  31r (26r) " S 126r

form of citation occurs on four pages, where folio numbers are listed under classified titles of machines.[9] It also seems evident that Taccola kept various books or fascicules; the separations between different groups of sheets are visible at the seams of the bound volume.[10]

The first authentically dated entries (I 42r, 61r) are of 1427.[11] The first of them is an unfinished letter copy, saying only *1427. Amicho karissimo*. The latter one of 1427 and those of 1438 and 1441 will be considered in connection with pages on which they appear, fols. 61r, 73r, 82r, and 96r. No other year is mentioned in the autographs until 1430 or 1431 (III 27r).

We conclude that Taccola, several years before 1427, began to write the contents of Lat. 197, probably using earlier drafts or versions, abbreviating some passages of the descriptions to "etc." In some cases he also entered drawings in some-

I 34r (29r)	at	S 126r
53r (49r)	"	III 29v
57v (54v)	"	S 126r
59v (56v)	"	S 126r
69v (66v)	"	S 126r
72r (69r)	"	III 32v
73v (69v, 70v)	"	III 32v
74r (70r, 71r)	"	III 57v
II 82v (79v)	"	III 29v
91v (92v)	"	II 81v
94v (95v, 96v)	"	III 32r

[9] The classified lists are entitled *De macchinis tormentis apprehandum castra et forte* (I 6v), *De currigi* (I 8v), *De bellis marinis et ullis rebus aque* (I 12v), and *De bonbardis* (I 13v).

[10] Especially after I 21 and before I 30. Other separations are visible at or near I 56, II 88, and S 118.

[11] The entry on fo. 42r otherwise contains only *1427*, the words *Amicho karissimo* at the head of the page, and drawings of piston pump systems.

what simplified or abbreviated forms for his own record. He kept a two-book unit, in which he initially recorded the *capitula* as a beginning part of the first book and the *Liber ignium* of Marcus Graecus as a beginning part of the second book. In the early 1430s, he then separately recorded Books III and IV (Palat. 766). Thereafter he added more than 500 Sketches in Lat. 197, along with smaller numbers of notes and classified lists of paginations leading to drawings, as well as full-page Addenda drawings, using blank spaces of Books I and II for Sketches and blank pages for Addenda. He added new fascicules of pages, and at some moment he repaginated the group of folios that are now a part of the Sequel. By these additions, Mariano converted his copy of Books I and II into the rather chaotic notebook it now is. Apparently he did this in preparation for his second treatise, *De machinis*, which he finished in 1449. By our tests of his handwriting and drawing technique, we believe that he probably added to the Sequel until the end of his life. It seems improbable that he would have so degraded his manuscript if some other copy of Books I and II had not existed, and as already mentioned, there is evidence in his use of "etc." that such a copy once existed. Copies or partial copies were probably made by the friend who is addressed in the note of 1427 (I 42r) and by those who saw all or part of the book in 1438 and 1441 (I 73r, II 82r, 96r).

The manuscript has no front page or title, but it does have the *incipit* of Book II (*Liber secundus*), which is the dragon emblem on folio 76r as now numbered (Pl. 13). The codex is not signed or dated on the page that is now near the end, but Taccola's signature appears at II 96r, near the end of the material that he designates as forming his Books I and II. On this page, under date of 15 August 1441, he wrote and signed a note stating that a Spanish priest saw the entire work, including its earliest part:

1441. Dominus Antonius catelanus presbitere de catelonia de civitate Tortose die XV^a mense augusto vidit hic designis ac etiam rotulum in quo erant machine et tormenta antiqua designata ex manu mei Mariani Jacobi di Sena.[12]

As for Taccola's unsigned pages, their authenticity is adequately proved by his handwriting, which is homogeneous throughout Lat. 197, although the codex contains the work of a number of years. However, there are some extraneous additions to Lat. 197 at the end of the Sequel. These pages and a note in Book II are clues to the early history of the manuscript after Taccola's time.

One page (S 128r) and perhaps two others (S 129r, 130v) show drawings and the handwriting of Francesco di Giorgio.[13] Another more clearly traceable

addition appears at the bottom of an initial page in Book II (77r): *Jo. Alberti Widmestadii cog^{to} Lucretii Suevi 220. Munus Sanctii Librani.*[14] The life and career of Widmestadius (Widmanstetter, 1506–1557) are known to some extent. He was in Italy for long periods of time and often visited Siena.[15] The person who gave him the manuscript may well have done so in that city at some time between 1533 and 1542.[16] This donor may have belonged to the Order of S. Jacopo, as was true of Taccola and also of Widmanstetter.[17] After Widmanstetter's death, parts of his collection became the founding stock of the Mun-

---

[12] Mariano Taccola never signed in exactly the same form and apparently did not maintain exactly the same title or name for each book but continued to paraphrase and originate forms at each time. Conceivably the rotulus is the set of *capitula*.

[13] L. Michelini Tocci, "Disegni e appunti autografi di Francesco di Giorgio in un codice del Taccola," *Scritti di storia dell' arte in onore de Mario Salmi*, II, Rome, 1962, p. 203 and figs. 1–3. The handwriting on these folios of Lat. 197 compares with Francesco di Giorgio's letter of 1488 reproduced in G. Mancini, *Giorgio Vasari. Vite cinque annotate*, Florence, 1917, p. 56. The entries on these pages are copies or variants of drawings by Taccola,

brief descriptions, and computations. It is not known under what circumstances they came about.

[14] The entry "220" of the signature is written in Hebrew and represents the Hebrew number equivalent of the letters in his name (see H. Striedl, "Die Bücherei . . . ," p. 240). Such plays on words or numbers were popular among the minor Copernicans and bible students of the time and were sometimes imitated even by greater men, such as Kepler. As to the last three words of the inscription, they are hard to read and to interpret, since the word *munus* may mean either work or gift. Widmanstetter usually wrote *donum* (see Max Müller, *Johann Albrecht Widmanstetter, 1502–1557*, Munich, 1907, pp. 24, 28, *passim*).

[15] As shown by Müller, Widmanstetter visited Siena in 1533, 1537, 1539, 1540, and 1542. He studied law under Marcello Biringuccio (who was related to Vannoccio, a follower of Taccola in engineering work) and under Alessandro Sozzini (descendant of one of the men who had shared excitement about military developments with Mariano, according to a note of 1438 at Book II 82r, quoted on p. 164.

[16] There was a family Santini in Siena (see Brandi, *Quattrocentisti senesi*, p. 229, regarding Jacobus Santini). In Bergamo, where a copy of Taccola's *De machinis* was sent to Colleone, the publisher Giovanni Santini is known (see Canestrini, *Arte militare meccanica medievale*, p. 33).

[17] Müller, *Widmanstetter*, p. 49.

ich Library; Lat. 197 was part of this stock.[18] When, in 1806, a new catalog of the library collection was made by Hardt, the manuscript was described by him for the first time as follows, in translation:[19] "197. Paper. Bound in wooden covers lined with white leather. Folio size. Written in a small, neat hand, with figures drawn with a pen. It once belonged to Jo. Alb. Widmestadius, as noted in his own hand, fol. 74 [77], lower margin; gift of Sanctius librarius. 128 sheets. Dated in the fifteenth century, as is clearly written, fol. 58 [61]. Moderately well preserved. Inscribed *Liber machinarum, et mechanica quaedam*[20] with brief texts in German and Latin. Among other things the *Liber ignium* occurs herein, fol. 74. . . . Greek historians attribute the invention of this fire to Callinicus of Heliopolis. . . . However, modern writers ascribe it to Marcus Gracchus; therefore, I believe that Graecus is in error for Gracchus. . . ."

Hardt, when citing folios in this description, uses Taccola's original folio numerals. In Hardt's total count of folios, apparently only Books I, II, and the Sequel—the latter exclusive of the blank sheet now counted as 137—were considered part of the *Liber machinarum* of 128 sheets. In fact, the codex presently shows a modern pagination series ranging to 128, written by a hand which is distinctly different from the second of two modern paginations.[21] The 48 sheets of the Hussite Anonymus, with German text, were not listed by Hardt, but they and 8 sheets of the Separate Fascicule, without text, evidently were considered the *mechanica quaedam*. None of these 8 sheets has numerals belonging to the first of two modern paginations, but the ninth folio (31) now has the numeral 124 in addition to the first, indicating that its location in the codex was fifth from the end in Hardt's time. It remained at that place, where it is now 133. It may be concluded that folios 32–39 (now 23–29) of the Separate Fascicule then was directly at the end of the Hussite Anonymus, not in its present position.

The situation was the same sixty years later when librarians published a shortened extract of Hardt's description, announcing the codex for the first time in print.[22] Then, within another 20 years, historians became interested in the codex, and they also knew Taccola's manuscripts in other libraries. Jähns described the codex (1889), followed shortly by Berthelot (1891). Then the new edition of the printed catalog (1892)[23] recognized for the first time both the existence of 48 plus 136 sheets (although S 137 and II 92*bis* were disregarded) and the independent character of the Hussite Anonymus:

"I. 48 sheets. Ca. 1430. A book of machines (military and other) drawn in Germany, with some German explanations. . . .

"II. 136 sheets. Ca. 1427–1441. By Marianus Jacobi of Siena (see fol. 96; also called Taccola). A book of numerous military and other machines. . . . Copies of our precious codex are in Venice, dated 1449, and in Paris."

The second set of modern paginations in the codex was probably inserted between 1868 and 1889, and about this time the Separate Fascicule was misfiled at its present place.[24] When these developments are kept in mind, it becomes easier to understand the structure of Lat. 197 and the several layers of authentic work contained therein.

Palat. 766 in Biblioteca Nazionale, Florence, comprises Books III and IV of *De ingeneis*, complete with the *incipits* of these books on folios 28r, 58r. The work is executed in light sepia ink, which also appears in Lat. 197. It is free of inserted sketches and notes, as may be seen readily when comparing Plates 23 ff. with Plates 2, 7, 10, 14–17. It is also superior in the planning and execution of an unbroken sequence of illustrated pages, presenting a difficult subject matter with style and without monotony. (Fairly successful planning of individual pages may be found in Lat. 197, and there are a few short sequences, such as those seen in Plates 16–19, but no part of it can be compared, as a product of graphic and editorial arts, with Palat. 766 in such a sequence as Plates 21 to 108.) Taccola took care to achieve a new arrangement and a fine balance of drawings and texts, with a new device for interconnecting them in the form of zoomorphic vignettes. The book consists of two fascicules, each comprising 24 folios.[25] Its binding, a sheet of heavy vellum folded over the endpapers, is the original one. The manuscript is clearly dated and contains several signatures by Taccola. The handwriting is uniform, and so is

[22] C. Halm and G. Laubmann, *Catalogus codicum latinorum bibliothecae regiae monacensis*, I, 1, Munich, 1868, p. 31.

[23] Halm and Laubmann, *Catalogus; editio altera*, I, Munich, 1892, p. 41.

[24] See Striedl, "Die Bücherei," p. 231, about the rebinding of manuscripts during the nineteenth century.

[25] The first fascicule has folio numbers 27 to 50. The second has numbers 51 to 76, but the folded sheet comprising folios 55 and 72 (4 pages) is missing. It already was missing before Mariano Taccola wrote the book; his drawing on fo. 71v flows onto 73r. The authentic folio numbers in Palat. 766 are Roman numerals; we substitute Arabic ones for simpler citation of the four Books. Gentile (1885) proposed a modern pagination for Palat. 766; we disregard it because it would introduce unnecessary complications.

the drawing technique. Both elements are of types that also occur in Lat. 197. They are still more uniform in Palat. 766, the obvious reason being that this is the work of little more than one year, while Lat. 197 is the work of many years.

On many pages of Books III, *Liber tertius . . .* (28r) and IV, *Quartus liber . . .* (58r), are cross-references to its pages,[26] as well as explicit and accurate references to *Liber primus leonis* and *Liber secundus draconis,* traceable into the old pagination of Lat. 197. These citations make it certain that drawings on the folios cited existed there at the time of writing, 1432–1433. In this connection it is also interesting that the paper size of Palat. 766, measuring 30 × 22 cm is the same as that of Lat. 197. The water mark in Palat. 766 is of an Italian source similar to one in the first score of sheets of Book I, although the paper of Palat. 766 appears considerably older.[27] The matching and authentication of Palat. 766 and Lat. 197 seem almost perfect. Clearly they were intended as parts of a treatise in four books.[28] Mariano's au-

thentic dates in Palat. 766, and his biographic statements that refer to Sigismund, are clearly corroborated by entries in the German imperial *Regesta,* including the appointment given Mariano when the Emperor was in Siena and Mantua.

There is no evidence in Palat. 766 that Sigismund, to whom it is dedicated, received this particular manuscript, or accepted it, or made it part of any library of his. However, there is indirect evidence that Mariano did not keep it. While the condition of Lat. 197 suggests that a book that stayed with Taccola soon changed its aspect, Palat. 766 contains no significant additions of notes or insertion of sketches. There are minor supplements relating to woodwork (IV 74, 75) near the end of Palat. 766, apparently unrelated to his plan for the treatise. He made these drawings probably during a relatively short time that he retained the codex. Substantially nothing is added by persons other than Taccola.[29] Since several of the Quattrocento copies of Palat. 766 are of Sienese origin, we are inclined to think that the codex remained in Siena for some decades.

In modern times, Palat. 766 and the copybook, Palat. 767, appear together in Florence. This copybook, to be described later, belonged to the Strozzi family in the seventeenth century, and the autograph may also have belonged to this family of humanists. The Strozzi were

---

[26] III 27r		cited at IV 59v
33v–34r	''	IV 62v
36	''	III 37v
38	''	III 50v
38	''	IV 64v
39	''	IV 64v
40v–41r	''	III 44r
43	''	III 44r
52	''	III 53v
IV 62	''	IV 63v
66v–67r	''	III 52v

[27] Briquet, *Les filigranes,* no. 3272 (*Demicerf*).
[28] This capital fact has not been noted thus far. The nearest approach to it is the remark of Berthelot (1891), p. 475, referring mainly to the first twenty folios in Lat. 197: *Le tout, il semble, tiré d'un traité spécial rédigé par Mari-*

anus Jacobus, traité qui n'a pas été signalé jusqu'ici à ma connaissance.
[29] Except an inscription and computation in two lines, about *Una lunetta duno cantone,* added to fo. 76v by a Cinquecento hand. It could have been made by an architect-owner of the codex.

book collectors even before Taccola's time. We do not know exactly when or how these two manuscripts came to the Biblioteca Nazionale, and the catalogs do not specify.[30] Only unknown symbols appear on the inside covers and the first folio.[31] The contents were enumerated by Gentile (1885) and Jähns (1889) and were cited without detail by P. Fontana (1936), Thorndike (1955), and Gille (1964).

A few words must be added here with regard to Lat. 28800 in Bayerische Staatsbibliothek, Munich, the autograph of Taccola's *De machinis*, finished in 1449.[32] This manuscript is closely related to Mariano's earlier *De ingeneis* but not identical therewith. The handwriting is proof of his authorship, while the drawing style shows direct continuity with parts of the Sequel and the Separate Fascicule of Lat. 197. Also the contents are thematically related to the Sketches and Sequel, although none of the entries in either manuscript is a real copy of a corresponding entry in the other manuscript. Clearly, Taccola was too original for that, even in his old age. The contents of *De machinis*, the Sketches and Sequel, are rather different from Books I to IV of *De ingeneis*. On the other hand, straight-forward copying from Lat. 28800 may be seen in the work of a later generation. So far as engineering contents are concerned, this autograph is copied in substantially identical form in three codices: Spencer Ms 136, New York Public Library;[33] Lat. 2941 in Biblioteca Marciana, Venice;[34] Lat. 7239 in

[30] There is nothing about this matter in Gentile (1885) or in the unwieldy and unfinished compilation of F. Palermo, *I manoscritti Palatini di Firenze*, Florence, 1854.

[31] Folio 27r (Pl. 21) has initials MN, of unknown significance, which also appear in some other manuscripts of the Palatine collection. On the inside cover is written *N. nuovo 766*, and in pencil, *V 366*. On the back, written in red ink, is *E 5, 5, 280* and once more, *V 366*.

[32] It was formerly owned by Graf Daun, later by Graf Wilczek. Berthelot (1891) noted the manuscript but did not see it; Laborde (1924) described it with great confusion and published illustrations with erroneous identifications. In addition to exact agreement of its handwriting and drawing technique with Taccola's earlier work, there are signatures by Taccola. The date of completion in 1449 is not stated in the manuscript, which has lost some folios, but it is cited in the copy in Venice (Lat. 2941) and in the copy in New York (Spencer Ms 136), which contains a signature and date in August 1449. As to this manuscript and copies cited below, see Mariano Taccola, *De machinis,* ed. G. Scaglia, Wiesbaden, 1971.

[33] The codex was formerly in the possession of Andrea Tessier (1819–1896) in Venice, whose collection was sold in 1900 at auction (Jacques Rosenthal, *Catalogue des livres, mss etc provenants des collections de feu le chev. Andrea Tessier de Venice*, Munich, 1900). See also F. Donati, "Francesco di Giorgio Martini in Siena," *Bullettino senese*, IX, 1902, p. 159. This codex contains the copy of Taccola's preface, which we have quoted in translation on p. 17, and also has the signature of the copyist: *Marianus tachola alias Jachobi dictus archimedis de Senis ac conposuit ex propria manu de mense augusti anno curente 1449 finivit. Eimmo de borsatis designavit.* Taccola's name is retained by the copyist, Eimmo de Borsatis, in the text of fo. 85r copied from Lat. 28800, 80r.

[34] It was first described by Morelli (1776) and has been much considered since then, by Romagnoli (1830), Promis (1841), Napoleon (1846–1871), Valentinelli (1872), Canestrini (1946), and Thorndike (1955). Francesco di Giorgio is mentioned in it and may well have added to it. Reti (1964) designates it as a "holograph" by Taccola. We cannot agree, if he means an autograph. It is sufficient to inspect the facsimiles in Canestrini (1946) and Berthelot (1891) to notice at once that radically different writings are involved.

Paris, Bibliothèque Nationale,[35] an elaborately produced, brazenly plagiarized copy by Paolo Santini. These *De machinis* copies are of interest in problems of *De ingeneis* only as background material for consideration of Mariano's style and technical concepts.

[35] It was first used by Carpentier (1766), then described by G. B. Venturi (1797, 1812, 1815) and Marsand (1835–1838), and subsequently considered by the historians cited in note 34 and by Berthelot (1891). It is an identical copy of Lat. 28800, wherein Santini substitutes his name for Taccola's in textual references. The figure style indicates that it may have been executed ca. 1450.

SELECTED DRAWINGS OF
BOOK I AND THEIR
HISTORIC POSITION

2 The opening part of Book I comprises illustrated texts, which Taccola repeatedly calls *capitula*. These Chapters occupy folios 1 to 21, and they then continue briefly on folio 30 and perhaps 31 (Pls. 1 to 6), after an interruption caused by insertion of a much later Separate Fascicule (Pl. 113). Most of the *capitula* occupy one page or about one-half of a page. They are closely interconnected and show that planning preceded the setting down of the existing formulation. The text of one Chapter repeatedly refers to the contents of another, citing accurately its title or opening phrase.[36] The entire group seems intended as a set of brief specifications, economically fitted together for the information of readers interested in clear, fundamental information, not in rhetoric. The contents relate to harbors and their construction, operation, and defense. No earlier collection of this kind—either in form or content—is known, except that Philon of Byzantium is said to have written *Limenopoiica* (On harbors), which was lost in Roman antiquity. Perhaps Mariano's work was based on medieval recollections of such a prototype.

Taccola's execution of the *capitula* may be only a draft, carrying out a complete plan in incomplete fashion. In fact it must have been slightly imperfect from the start, as several folios, such as I 1, 7, and 8, have their verso side upside down when the recto is in normal position. Some entries, for instance at I 2, 3, and 4, are substantially less clear than other writings by Taccola; they may be some of his earliest exercises in technical writing. The abbreviation "etc.," less frequently used in his other writings, is often found in the *capitula*, and sometimes the meaning is uncertain. The "reader," whom Mariano addresses in these Chapters, could hardly be expected to supply details indicated by this abbreviation.

A clue for dating the Chapters may be provided by an introductory note, I 1r:

Take note of Christmas Night. On that night it is settled for the coming year whether there will be wealth or poverty. That is, if the moon then is only one day old, the gain of the day will be poor. The more she has grown by that night the greater is the gain. If she is altogether full, there will be great fertility, etc.

Taccola seems to ponder an important enterprise, perhaps the beginning of this book. He becomes hopeful of a prosperous astrological omen, full moon on Christmas night. During a major part of his adult life, a constellation approaching this position occurred only a few times: in 1417, 1419, 1427, and 1430.[37] The first of these or any earlier date would be improbable for a beginning of his book, as ten years would have elapsed until his next dated entries; such extended silence would be out of character by

---

[36] This was first noted by Thorndike (1955), p. 18.

[37] Computed from the data in B. Tuckerman, *Planetary, Lunar and Solar Positions* A.D. *2 to* A.D. *1649*, Philadelphia, 1964.

standards of subsequent contents of the book. The last of the astronomically fixed years is also improbable, since by then his Third Book already was in process. It seems that he planned and began the Chapters in either 1419 or 1427. The latter year may be questioned because about that time he already spoke of his efforts over an extended earlier period (I 31r).

Some of Taccola's descriptions deal with structures that others take for granted but which he presents as a kind of large-scale engine. This is true for instance, of his Chapter on Good Harbors, I 13r (Pl. 1).[38] The picture shows the diagram of a harbor in use in Taccola's time. He writes,

> Good harbors must be narrow at the entrance and large inside. They must be deep. At the entrance always provide fortresses or guard towers. At this place let an iron chain be provided, which can be drawn in at night by a wheel or drum. Make it long enough that it can go down into the water. If the chains were not [so] long, they would be use less because [your own] ships would run into the chains as they pass, etc. When harbors are fortified as shown above, the enemy cannot attack the ships in the harbor, etc.[39]

This procedure was in keeping with actual practice. It is doubtful that the practice was described in earlier books of the Middle Ages, since Mariano's description and illustration seem to be of a tentative and groping character. The "wheel or drum" is not shown. On the preceding and following pages, which are more defaced by sketches than this one, Taccola shows a harbor with a chain of greater complexity (12r) and details of still other chains (13v). When describing these details, he is careful to refer back to 13r by saying that the chain "may be drawn . . . as noted in the Chapter ¶ 'Good Harbors' etc." This is one of the cross-references typical of the *capitula*.

On other pages, showing engines of naval war, Taccola makes further specific reference to the defense of harbors. These pages are distributed between I 1v and 20r. On a dozen of them he shows some thirty types of underwater defenses, technically known as Chevaux de Frise. Some of them can be connected to harbor chains and raised and lowered with them, or boats can be lifted over them.[40] More briefly he deals with the well-known grappling hooks, ship's drills, landing bridges, fire throwers, and incendiary bombs. These were illustrated in previous books,[41] but some of

[38] Copies of this drawing are in Urb. Lat. 1757, 30v; Add. Ms 34113, 103v. Later variants are in S IV 1, 109v, 111r, 112r.

[39] Porto Pisano had iron chains of this kind between its *torri delle catene*. The Florentines removed them when they acquired the citadel, city, and harbors of Pisa (1405, 1406, 1421), but the chains were later returned and can now be seen in the Campo Santo of Pisa. Meanwhile the old harbor silted up, and the remains of the towers are now far inland.

[40] Floating harbor chains in the *capitula*: 13v (copied by Berthelot, 1891, fig. 26). Bottom-mounted poles with top members passable only in one direction: 12r. Barrels tied to a boat to lift it: 6v, 11v (Frontinus, *Stratagemata* I, 7); also in Guido da Vigevano, see Berthelot, 1900, p. 416 ff. Barrel-supported walkway: S 115v; Vegetius, III 7; Frontinus, IV 28.

[41] See F. Miltner in Pauly-Wissowa, *Real-Encyclopaedie*, Supplement V, Stuttgart, 1931, s.v. *Seekrieg*, cols. 898–905. Copies of Taccola's drawings occur in Urb. Lat. 1757, 18r to 30v; Add. Ms 34113, 104v to 109r, 117r to 125r; Uffizi drawings, Arch. 1450, 3976, 4062, 4068, S IV 1, 108v–109r, 114v–115r.

Mariano's versions are very complex and may be new proposals. Devices of this type also reappear in later works, including Alberti's *De re aedificatoria* (IV, 8 and V, 12). Strangely, Alberti mentions them without detail or illustration but claims improvement and original thought, while Taccola shows and describes mechanical gadgets of many types but makes no claim of this kind. The impression arises that the engines of war and defense were a matter of popular interest, that Taccola developed graphic forms for illustrating them, and that Alberti developed a personal, literary style for their description.

Sheet 14r introduces the cannon or bombard.[42] It had been described in the earlier *Bellifortis* by Kyeser and the still earlier *Feuerwerksbuch*.[43] At this point Taccola comes closer to an explicit discussion of new inventions:

. . . Bombards of iron. It is well to make them of iron, but is best to make them of cast iron (*de ferro colato*) since this is of tough nature and makes harder tubes. When made of pewter, they are not durable but break rapidly. Bronze is better, using pure copper, the cast, etc., of which is infrangible when shaped as shown, although it appears ductile and easy to fabricate. . . . Note that bombards of great weight cannot be carried on carriages. . . . For this reason bombards of two and three parts have been invented. . . . The gaskets of these bombards must be of soft wood, in order that the tubes may be sealed together more closely, etc.

These bombards of two or three parts—not a very successful invention, as it turned out—appear in various sketches, for instance at I 21r (Pl. 3). Primitive cannon mounts are shown elsewhere, in sketches at II 82r (Pl. 14). Various gunpowder mixtures are prescribed, on folio 14r in a subsequent insert. However, the real description of explosives, among other chemicals, comes in the *Liber ignium* in Book II.

Not everything that is in the *capitula* would be classified as a harbor construction or as a weapon, according to modern concepts. For instance at I 15r (Pl. 2) appears the drawing of a bucket well.[44] These of course are of biblical and prehistoric origin and were in daily use. They were discussed in Chapter X of Book IV by Vegetius on the defense of cities, *ne aquae inopiam patiantur obsessi*. Taccola seems to include them in the same spirit, and the same practice was followed in military books at least for a century after his time.

While probably considering wells at this point only as weapons of defense, Taccola is careful to consider them thoroughly. He shows a variety of mechanisms, I 15v. 19r, 19v,[45] and he later shows them in what, for his time, were more advanced forms, for instance at 20v.[46] Then at III 47v, he refers back to

---

[42] Copied in Urb. Lat. 1757, 42r; Add. Ms 34113, 130v; Palat. 767, p. 157; S IV 5, 10v; also copied in Berthelot (1891), fig. 27.

[43] Berthelot (1900), pp. 392–396; Jähns, *Geschichte*, pp. 229–236, 382–408. There are still earlier Byzantine and Arabic antecedents (Jähns, pp. 156 ff, 181 f).

[44] Urb. Lat. 1757, 34r. As to Hellenistic-Islamic antecedents, see E. Wiedemann and F. Hauser, "Über Vorrichtungen zum Heben von Wasser in der islamischen Welt," *Beiträge zur Geschichte der Technik und Industrie*, VIII, 1918, pp. 130–132, 151.

[45] Copied in Urb. Lat. 1757, 34v, 35r, 35v.

[46] With crude representation of a chain wheel actuating a spur gear drive. Copied in Urb. Lat. 1757, 36r. Published frequently, for instance by Berthelot (1891), Fig. 30, and Beck

the present chapter (under its original folio number 16). The chapter on I 15r by itself discloses only that the pulley should be well rounded and balanced and equipped with handles; the text says, "Make it of *** foot height," but the dimension is omitted. As to dimensions and similar details, Mariano states here, "Not everything that is in these designs must be said expressly, etc."[47] The principal object here in the *capitula* is to develop an arsenal of machines. He develops his "Theater of Machines" before our eyes and claims that he knows more than what he shows and specifies. Apparently he sought employment as an expert.

Mariano shows two types of Double Bellows Systems, for the continuous supply of water or air, at I 30v and 31v (Pls. 5, 7). In the first of these drawings he shows a water-driven camshaft in front, two large bellows in the middle ground, and a furnace in the background, and he uses sharp perspective foreshortening—a technique that is unusual in his drawings. The bellows system is one variant of a famous development that gradually brought the blast furnace and its product cast iron. Did Taccola know the exceptional importance of the unit illustrated here, and did he illustrate it in unusual perspective for this reason? We can hardly believe it.

The development of perspective began centuries before Taccola and continued centuries after him. However, he surely applied great care in developing proper forms of showing bellows, and his book contains a number of studies of these devices (I 53r, 53v, 61v, 63r).

The illustration at 31v shows foot-operated, water-pumping bellows. The foot-operated, air-blowing variant had been known already to the ancient Egyptians. Even in developing this age-old system, Taccola uses considerable effort. At some later time he added a description and related suction pipe detail, in a later form of his hand, and it may have been then that he added the human figure operating the machine. Probably still later he added Sketches at the bottom of folios 30v and 31v, including the citadel on the latter page inscribed *Tolleti*.

One of Taccola's manifold interests is the practical side of architecture, the work of the building superintendent. In the *capitula* he shows tools used by carpenters and blacksmiths of his time (I 16v–18v). Here he also includes equipment for divers who descend into water to work on harbor chains.[48] He also includes door hinges. Elsewhere in Book I he shows roofs, 37r; chimneys, 32r ff (Pl. 8);[49] and stairs, 73r f. A house front,

---

(1899), Fig. 346. Spur gears are otherwise lacking in Taccola's work, while chain drives are extremely rare. The unique drawing in Book I may indicate that Taccola, in his study of well mechanisms once observed the device of 20v and made this record of it. He did not repeat it in his *De machinis*.

[47] In I 15r· *Et nota semper quidquid designatur in eo non sprimitur.*

[48] As noted by Berthelot (1900), pp. 106–113, the collection comes from earlier precedent, also reflected by Kyeser, *Bellifortis*, fols. 125 ff. Most of the tools, as well as most of the ladders, rope ladders, and the like, are copied closely in the later notebooks, while being varied in Books III and IV and *De machinis*. For identification of the copybooks, see pp. 191 to 203.

[49] See one in Pl. 8. Another (I 32r) was reproduced by Berthelot (1891), fig. 37.

shown in one of the copybooks and collections of notes (Cod. S IV 6, 34r) may also derive from architectural work, either done or studied by Mariano. At least one of the architectural folios, I 18 (20), is cited later (II 92 *bis* r; IV 62v).

The *capitula* texts contain other ideas pertaining to warfare. Mariano shows a mounted gunner at 21r and an ancient stratagem on 21v (Pls. 3, 4).[50] The stratagem is well known in the literature (Frontinus I, 7, 8; Kyeser, *Bellifortis*, 87r); here Taccola describes it in terms of unusual prolixity, compared with his general style. Similarly prolix is the textual description of the gunner in this chapter. The gunner is repeatedly called *bobbardarius*, while other pages designate the weapon carried by him as *bombarda*. Folio 21r seems to be copied from an extraneous text.

Bizarre elements are introduced on the last folios of the *capitula*, where Taccola shows fantastic divers, one riding an ox and the other a fish. The ox rider (I 30r) says, "I keep to myself what I know. Do not fancy that you can do things without me, the silver-bearing ass, etc."[51]—a rather direct statement that the inventor's "secrets" are available only for money. The fish rider (I 31r, Pl. 6) begins with a joke or riddle and ends on a note of protest:

I know what I am doing on the swimming fish. I feed him oil from the sponge, as spittle for the fish, while riding him. He has in himself what supports him and what is supported by him. ¶ Nothing in the *capitula* on harbor gear and defenses can be brought to perfection without me. My speech is veiled.[52] Many a thing that I have acquired with long labor shall not be known at once. What I say here I say because of the ingratitude of the people, who are less than human. The rest reposes in my mind, etc.

Evidently the divers speak for the author. They are also used as symbols; in fact the one on 31r has the head of still another diver as an emblem on his shield. Taccola's full intention in creating these images may escape us, but one meaning seems to be that the diving figures symbolize his interest in harbors and hydraulic construction. Elsewhere, when showing diving equipment, he illustrates what appear to be matters of technical interest (I 12r, 57r, III 44r), as did his contemporaries Kyeser, the Hussite author, Fontana, and others. Those men were fascinated also with old legends about flying engines; Taccola is silent about them.[53] His diving scenes, as already noted, are symbols for unusual work, not only direct illustrations of observed or proposed work.

After the *capitula* Taccola continues to concern himself mainly with hydraulic

[50] Copied in Add. Ms 34113, 130r.

[51] The statement is perhaps an allusion to Frontinus, *Stratagemata*, III, 1 (8, 9). Its illustration is copied in Urb. Lat. 1757, 33r; Add. Ms 34113, 72v; Uffizi, Arch. 3976; Berthelot (1891), fig. 35.

[52] *Velate locutus sum.* Thorndike (1955), p. 18, tried to read this text but mistook *velate* for *relate*, also *equitans* for *equitonis*.

[53] For ancient diving operations, see Daremberg-Saglio, *Dictionnaire*, s.v. *uter, follis*. For Kyeser, *Bellifortis*, 55r, 62r, see Berthelot (1900), pp. 345, 353. The well-known legends about attempts to fly are as absent from *De machinis* and the copybooks as from *De ingeneis*. Reti (1964), considering only a single drawing, I 71v, interprets a *puerorum ludus* by Taccola as a helicopter, in fact one with possible military application. The theory appears based on misconception of Taccola's interests and methods.

constructions.[54] The first twenty-one
folios are occupied by the Chapters,
another eight by the Separate Fascicule
of later times. Then Book I continues
with forty folios or eighty pages, at
least fifty pages of which show some
kind of waterworks: harbor structures,
machines for building them, means to
raise water, and machines such as that
of I 40r (Pl. 10), possibly a fulling mill
driven by a waterwheel and camshaft.
The drawings are in very different stages
of completion in this part of the book.
Rough, preliminary and exploratory
sketching is frequent, and texts are rare.
There are no more cross-linking references that would show the continued
presence of a planned work.[55] In this
area of the book, which seems to contain
outlines for hydraulic construction, Taccola claims inventions of his own (I 61r).
He writes after one of his rare invoca-
tions of divine assistance,

Be it noted that in the year of the Lord
1427, on the 6th day of December, we
completed four inventions and finished
their test. ¶ First, that bridge piers can
be founded in the Tiber, without de-
flection of the river and without suction
pumps for the caisson. ¶ Second, that
breakwater walls in the harbor of Genoa
can be built and founded when time is
scarce. ¶ Third, that gristmills can be
constructed on lakes, without incline,
in Gallic or country style, as desired by
the miller. ¶ Fourth, that a caisson can
be kept empty, without use of buckets
and water pumps. [It can be] 40 foot
high and sixteen or twenty-four foot
wide, with a bottom in the lower part
and none in the upper, and provided
with an aperture in the bottom. It is
submerged. The aperture can be as large
as the head of a man. Again: it remains
empty, etc.

These inventions are illustrated in
later chapters in Book III (Pls. 34, 50). In
Book I, folios 44 ff, Taccola shows a
variety of elements such as caisson details and methods of measuring the
ground for their use, as well as various
water wheels, piston pumps, bellows,
and other devices usable for removing
water from the space within a cofferdam or for supplying water to a city.
He also shows more comprehensive
systems. In addition, he shows a mercury-driven mill, I 74r, apparently a
model or demonstration device. Mercury drives were known in clocks.[56]
Taccola's page discusses only mechanical
details: "Mill for milling by means of
mercury. Make the bearings of cedar or
orange wood. . . . Young laurel is also
good." Part of the short inscription is
illegible.

[54] Regarding I 35v, 36r, bucket wheels, and 54r,
duct work, see Berthelot (1891), figs. 38, 39, 45.
[55] Nevertheless, many of the drawings are
copied exactly in Urb. Lat. 1757, 3–15, 55–60,
in Add. Ms 34113, and in drawings in the
Uffizi attributed to Antonio da San Gallo the
Younger; close versions are shown in Palat.
767 and S IV 5. The publications of Beck,
Feldhaus, Forti, White, and others seem to
take particular interest in a windmill, I 87r,
and a *perpetuum mobile*, 58r.

[56] S. A. Bedini, "The Compartmented Cylin-
drical Clepsydra," *Technology and Culture*,
III, 1962, pp. 115–141, especially p. 116. Mercury drives were also proposed for perpetual
motion machines (Villard de Honnecourt,
Bauhüttenbuch, ed. H. R. Hahnloser, Vienna,
1935, Pl. 9, copied for example in F. Klemm,
A History of Western Technology, Cambridge,
Mass., 1964, p. 90). Taccola's model may be
considered a distant forerunner of the more
basic, quantitative, hydraulic models, tested
by G. Poleni and G. B. Venturi, ca. 1718–1818
(see H. Rouse and S. Ince, *History of Hydraulics*, New York, 1957, pp. 113, 136).

1. Harbor structure **(I, 13r).**

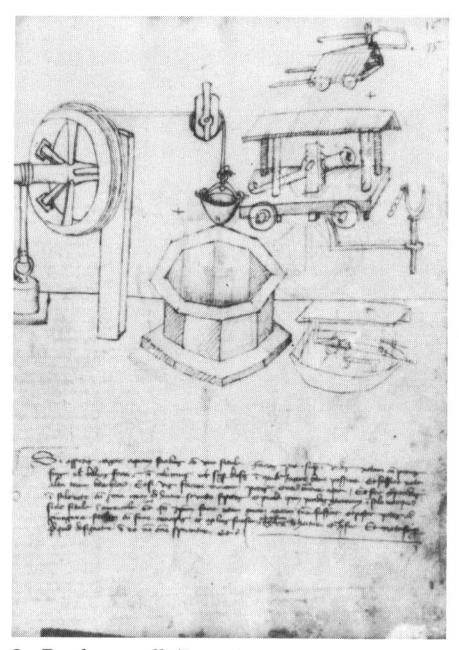

2. Bucket well **(I, 15r).**

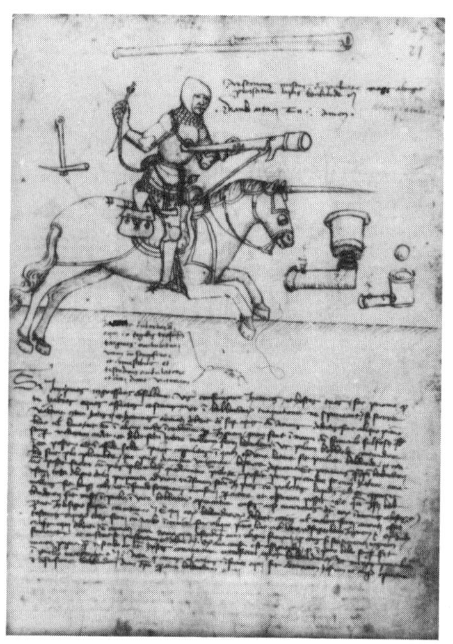

3. Mounted gunner **(I, 21r).**

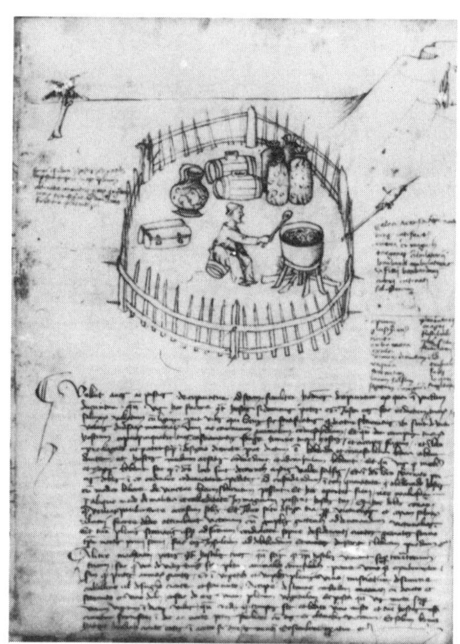

4. Stratagem **(I, 21v).**

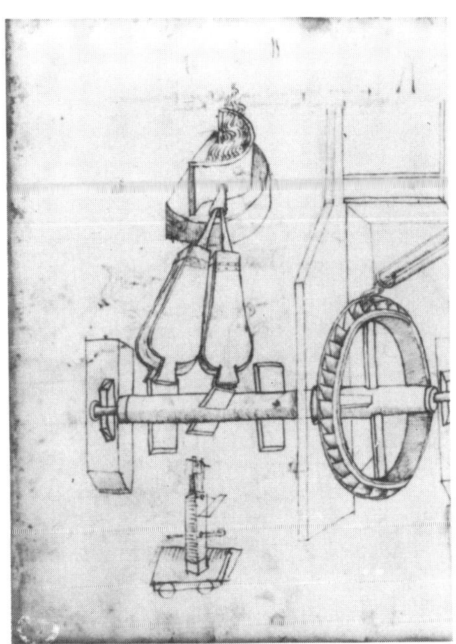

5. Double bellows for air blast **(I, 30v)**.

6. Diver **(I, 31r)**.

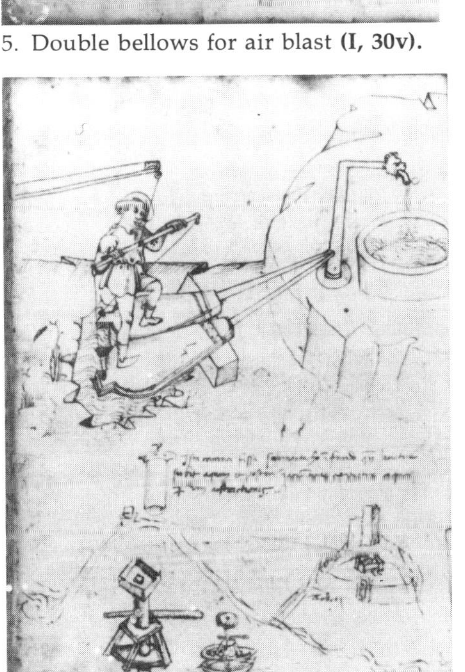

7. Double bellows for a pump **(I, 31v)**.

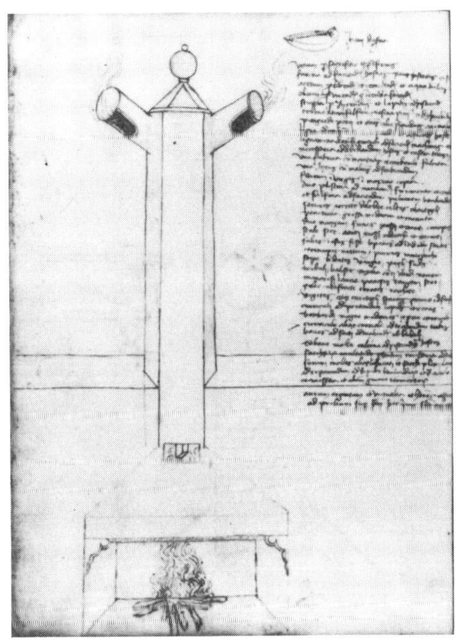

8. Chimney **(I, 34v)**.

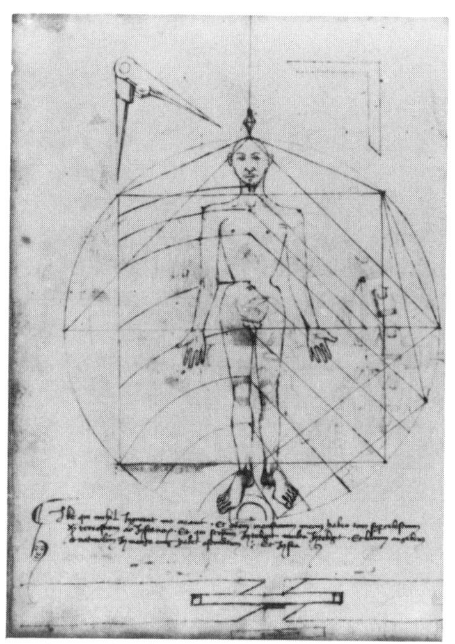

9. Microcosm man **(I, 36v)**.

10. Fulling mill (?) **(I, 40r)**.

11. Man starting a siphon, and sketches **(I, 68r)**.

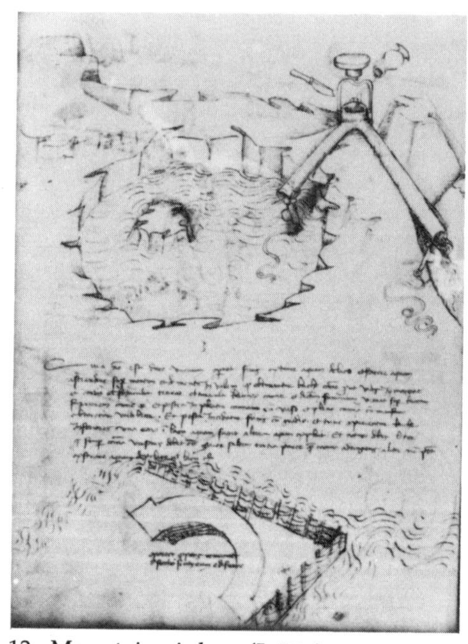

12. Mountain siphon **(I, 73v)**.

SELECTED DRAWINGS OF BOOK II AND THEIR HISTORIC POSITION

3 The first page of the second book (II 76r, Pl. 13) shows the emblematic dragon with which Taccola identifies this book when citing its contents in subsequent books. The title above the dragon is *Liber secundus*. The introductory text on this page reads, in part: "Be it noted that a leader who wishes to acquire some country or province must mainly and principally have the country . . . mapped . . . , in order that he may be better informed. . . ." Aphorisms of this kind, no doubt derived from Frontinus and Vegetius, are collected in greater number in Mariano's *De machinis*.[57] Some of them may have appeared on the title page of his *Liber primus leonis*, now missing. The title page and introductory part of the authentic *De machinis* (Lat. 28800) is also missing.

Book II then devotes five narrowly written pages (folios 77r–79r) to a subject that we may call medieval chemistry. It mainly comprises Mariano's transcription and partial modification of the famous or infamous book, explicitly credited by him to Marcus Graecus, *Liber ignium ad comburendos hostes*. This early Byzantine work had previously

been copied and developed by many others,[58] and at this point it is closely copied and slightly developed by Taccola. He adds recipes for ink and the like, which he probably takes from contemporary or slightly earlier manuscripts of other professional writers.[59] As already indicated by Berthelot (1893), we have here a developing tradition, the beginning of which goes back to Hellenistic times. There are glosses by Roman, Byzantine, Islamic, and medieval Latin copyists, often reflecting only semantic differences but sometimes adding information based on new experience or experiment. As the latest copyist's additions often come at the end of a compilation, it is here that Taccola's additions to the book of Marcus Graecus are found in Lat. 197. His version of the medieval chemical instructions or recipes then was copied by others, along with his illustrations of cannons, forges,

---

[57] These aphorisms are found in Spencer Ms 136, fols. 2r to 6v, and in Lat. 7239, 5r to 9v.

[58] Hardt (1806), p. 514; Berthelot (1893), pp. 100–133; Thorndike, *History of Magic*, II, pp. 738, 784 ff. Also see note 43 *supra*, and Berthelot (1895), *passim*.

[59] II 79r: *Faciendum actrementum accipe aque pluvialis (uncias) 15 et micte in ollam novam et fac in uno baculo signum mensure (quantitatis?) aque et postea adde tandem aque . . . et postea adde 30 unciis aque . . . galle piste (tertios sex?) et fac bullire usque ad . . . signum baculi et postea adde gumis (uncias) 3 (et dimidium) et fac bulire usque ad consumandam aquam . . . semper duciendo eam et postea cola deinde desuper (vi)t(r)-eolum ad quantitatem (dramma 4?) cum vini albi libre 3 et cum fuerit refrigidata . . . micte per unam noctem in sereno et usui reserva etc.* The same procedure, in partly the same words except for slight variation as to ounces and pound signs, is specified in the instructions written by Albertus Porzellus in Milan, prior to 1382, copied and slightly restyled by Taccola's contemporaries in France: M. P. Merrifield, *Original Treatises on the Arts of Painting*, I, London, 1849; New York, 1967, pp. 288–291.

bellows, and hydraulic machines. Chemistry was developed further and more significantly in the *Pirotechnia* of another Sienese, Vannoccio Biringuccio, posthumously printed in 1540. Biringuccio writes on folio 164v, "This information is [derived] from a little work that came into my hands long ago. It had been very anciently written on parchment." Folios 77–79 of Lat. 197 may be Taccola's private copy of this parchment codex.[60]

Book II contains further hydraulic-mechanical drawings, such as piston pumps, bucket wells, and siphons. There are also military Sketches of warriors and stratagems, ranging to folio 96.[61] The combination of objects is complex, and at some points confusing, as may be noted in II 82r, 82v (Pls. 14, 15). The verso of this sheet contains the schematic drawing of a piston pump, particularly a suction pump,[62] which Taccola cites in a subsequent part of *De ingen-*

eis, III 29v. At later times, as indicated by style analysis, he added a small drawing at the bottom of the sheet, schematically showing a water supply system, and he also added Sketches of a cannon, a cannon mount, and a cannon transporting team. Even more confused is the recto, which originally showed in the center of the page a cistern and a short text about it. Later the page was filled with minute sketches and notes, one of which states that the cannon aiming devices, included here, were shown to Piero de Micheglis on 9 December 1438. He is the councillor who was named *comes palatinus* in Mantua, 1433, when Mariano was also honored.[63] Taccola's note on this page implies that *Messer* Piero was impressed: "He then said that he wanted at once to confer with an aide of [the condottiere] Francesco Picc[inn]ini." Another note refers to the preceding day: "*Messer* Mariano Sozzini saw all these [drawings] on the 8th day of December in the house where he resides."[64] On an earlier page (I 73r) is a similar and pertinent note, dated a few weeks earlier, saying that Daniele Nicolai de Romanellis was told about the invention of two-piece and three-piece cannons.[65]

---

[60] Substantially exact copies of Taccola's modified copy of the Marcus Graecus book are in Urb. Lat. 1757, 219v–235r; Add. Ms 34113, 18v–27v; and (incomplete) Palat. 767, 250–252.
[61] Many of the devices and some of the drawing compositions are known from earlier work. Floating belts, II 90v, 91r, are shown by Kyeser in very similar form (see Berthelot, 1900, pp. 355, 357 and previously by Guido da Vigevano, *Thexaurus* (unpublished). Relatively little of this material is in the copybooks, but substantial portions of Book II are in Add. Ms 34113.
[62] Reproduced in White, *Medieval Technology*, Pl. 7, and plausibly discussed at p. 113. An assumed problem, predicated on this single drawing, is discussed by J. Needham, "The Pre-Natal History of the Steam Engine," *Transactions*, Newcomen Society, XXXV, 1962–1963, p. 45. The problem disappears when *De ingeneis* is considered in its entirety. Also see Wiedemann-Hauser, "Heben von Wasser in der islamischen Welt," pp. 145–149.

[63] *Supra,* p. 16.
[64] This was Mariano Sozzini the Elder (1397–1467), teacher of Canon law at the *Studio,* rector of San Pietro a Ovile, and author of various books posthumously published, for example *Commentaria in ius canonicum,* Venice, 1593; *Viaggio in terra santa,* Florence, 1822; and "De sortilegiis," *Archivio per le tradizioni popolari,* XV, 1896, pp. 1–7. Concerning Sozzini and Pietro de Micheglis, see Zdekauer, *Studio,* pp. 164, 167 f.
[65] *Die 3a mensis septenbris hanc bonbardam anno 1438 indicavi danielli nicholaii romanellis de Sena.* Daniello Nicolo Romanelli is otherwise

Evidently Taccola submitted his work to
the leaders of Siena. Sigismund had
died in the previous year.

Taccola's last dated disclosure is on a
folio with earlier drawings: "1441. The
Reverend Antonius of Catalonia, pres-
byter of Tortosa, on the 15th day of
August saw the designs entered here and
also the booklet of ancient machines and
weapons, designed by me, Marianus
Jacobi of Siena." The "booklet" (*rotulus
in quo erant machine et tormenta antiqua*)
may have been the manuscript of the
capitula or some lost writing. The purpose of showing the war machines to the
man of God is not stated. Perhaps Taccola wished to obtain some kind of
official registration for the *De ingeneis*.

known only through his letters of 1423, written in Rome (Valois, *Le pape*, I, pp. 14 n. 1,
16 n. 1).

13. *Liber Secundus.* Dragon emblem
(II, 76r).

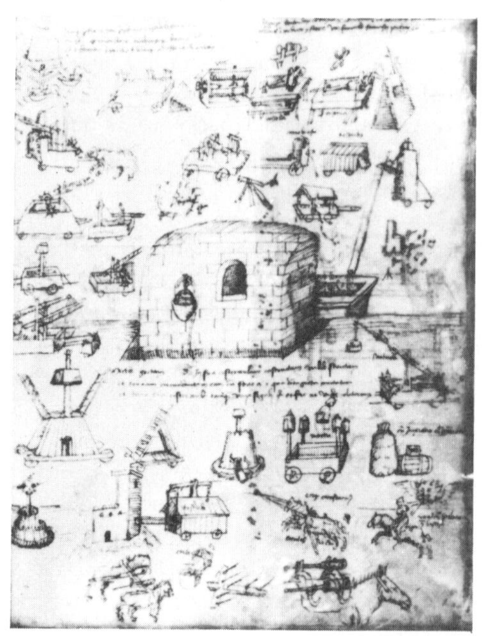

14. Cistern and military sketches **(II, 82r).**

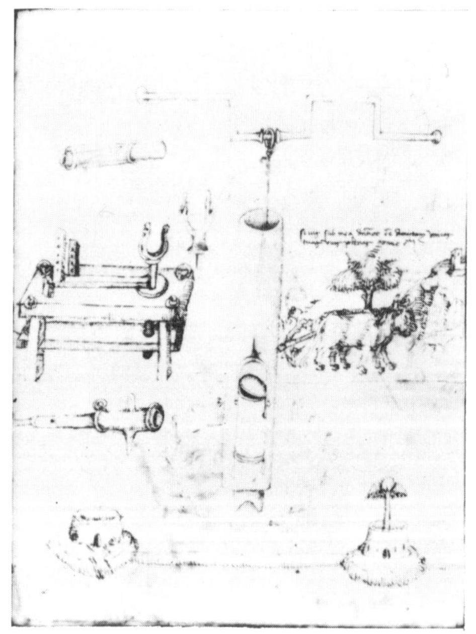

15. Piston pump and sketches **(II, 82v).**

16. Rider floating in water **(II, 90v)**.

17. Soldier floating in water **(II, 91r)**.

18. Bridge siphon **(II, 94v)**.

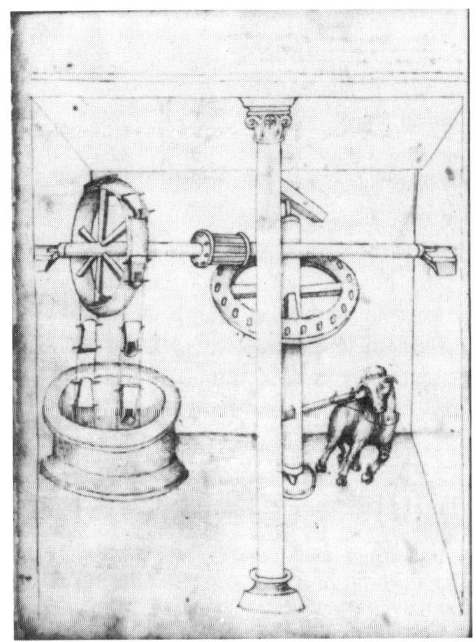

19. Ox working a gin **(II, 96v)**.

THE DRAWINGS AND TEXTS
OF BOOK III AND THEIR
HISTORIC POSITION

4 Book III begins with two portrait drawings (Pls. 20, 22) and then presents drawings of machines and technical methods. Taccola's texts usually describe the illustration and identify it by a figurative vignette on the margin next to the descriptive text. A distinct and original form of balancing the drawing, text, and vignette is used, as may be noted, for instance, from Plate 34 or Plates 44, 45. The handwriting is always the same as shown here. It begins near the upper edge of the verso, except where a double-page format is used, and the vignette is always at or near the left edge of the text.

Taccola's texts in Latin and then English translations are to be found with the plates in this volume.

| III 27r. Mariano's Preface to Book III. Illustration of "mensura." (Plate 21).[66]

The book begins with an invocation and a date. This date (MCCCCXXX.) appears to have four dots below its last digits, actually the ends of downstrokes in light ink. It may once have read MCCCCXXXI. The writing is defaced at several places, including the line di-

rectly following the date. If Taccola wrote "1431," it becomes relatively clear why he also speaks of an "arrival" of Emperor-elect Sigismund. It was in the autumn of 1431 that the prince reached Italy. He then came to Tuscany in the early part of 1432, possibly before 24 March, the end of the Annunciation year 1431, and to Siena in July of 1432.

The page continues with a somewhat enigmatic preface, wherein the writer mixes notarial formalism with stilted allusions and highly personal forms.[67] Abbreviations are used in extraordinary profusion. He even abbreviates his own baptismal name to a capital M. If our reading of the difficult text is correct, Taccola first states, with obvious self-assurance, that the contents of the book have been shown to the visiting Emperor-elect in July of 1432, and he then adds an *apologia*. He develops very much his own version of a disclaimer: on the frontispiece, III 27v (Pl. 22), God admonishes the king to protect his people, while here on the recto it is considered that Mariano in turn may be admonished, *increpari*, by the king to become more proficient than he is. It is unknown whether the unusual terms refer to some particular discussion. The impression arises that a dedication (IV 69v) may have been written before Taccola's audience with Sigismund, as it speaks entirely in the future tense, but that the preface, using past tense, was inserted

[66] Copied in Add. Ms 34113, 22r. The frontispiece portrait, III 27v, never recurs in the copybooks, nor do the pictures of saints (III 69r, 67r).

[67] The titles that he gives Sigismund are almost literally those used by the king himself at this particular time. See *Monumenta conciliorum*, I, 1857, pp. 227, 265.

after the audience. Did Sigismund accept some copy of the book and take it with him? This one of Taccola's manuscripts, Palat. 766, remained in Italy.

Technically the preface emphasizes measurement by compasses, *cum circino mensuras*. Mariano had shown, as well as used, compasses in his Microcosm picture (Pl. 9). In Plate 21 he shows how to measure the height of a tower by an indirect method, presumably since the tower is in enemy hands, and one cannot ascend it and lower a measuring string. By some unspecified but obvious method, Mariano determines the distance from his ultimate location (at left on the drawing) to the foot of the tower. He marks it at a reduced scale on his drawing page, using his compass as dividers. He then determines the altitude angle of the top of the tower as seen at his location, by his quadrant; he no doubt graphically transfers it to his page, measures the resulting height of the tower, as shown on the page, and computes the actual height by using the predetermined scale.

| III 28r. Fountain. (Plate 23). A descriptive text, in 8 lines, and a duck vignette (Plate 24) are on the verso side.[68] The recto has the *incipit* of *liber tertius de ingeneis ac edifitiis non usitatis*, Book III, On Unusual Engines and Devices. With regard to this title, it is interesting that

Philon of Byzantium, in chapter III of his *Pneumatics*, cites his book On Extraordinary Instruments.[69]

The drawing presents the ornamental sculpture and the simple pipe system of a fountain—an application of the principle of communicating tubes, known from ancient practice and books, for instance Pliny (XXXI, xxxi, 412). This simple system may have been chosen to guide the reader gently into the technical subject matter. The need for "measurement," pointed out in the preface (III 27r), is alluded to by the showing of a leveling instrument. This A-shaped instrument was actually used in Siena's water-main practice.[70]

| III 29r. Piston Pump. (Plate 25). Text in 6 lines and stork vignette (Pl. 26) on the verso. The illustration is one of a few copied in the more selective copybooks, such as Urb. Lat. 1757.[71]

Although Taccola shows a piston pump as one of the "unusual engines," according to his chapter heading the basic principle was very old.[72] The valve ap-

<hr>

[68] Copied in Palat. 767, p. 20, with text on p. 237: *Una fonte daqua viva accio che getti lacqua alta bisognio . . .* The text is slightly different in Add. Ms 34113, 22v: *Una fonte a ffare che la gitti in nalto essendovi aqua viva . . . bisognia. . . .* A basically similar drawing had appeared in Kyeser, *Bellifortis*, 54r, but as shown by Berthelot (1900), p. 342, it had been confused

[69] M. Carra de Vaux, "Le livre des appareils pneumatiques et des machines hydrauliques par Philon de Byzance," *Notices et extraits des manuscrits de la Bibliothèque Nationale*, XXXVIII, Paris, 1903, p. 124. Similar titles occur in the *Paradoxographi graeci*, for which see A. Westermann, *Paradoxographi graeci*, Braunschweig, 1839, and O. Keller, *Rerum naturalium script. graec. minores*, 1877.

[70] Bargagli-Petrucci, *Le fonti*, I, p. 42; Daremberg-Saglio, *Dictionnaire*, s.v., *diabetes* and *chorobates*.

[71] Urb. Lat. 1757, 8v; it also is in Palat. 767, p. 32, text p. 237; Add. Ms 34113, 23r.

[72] Beck, *Beiträge*, p. 288 ff. Also see Daremberg-Saglio, *Dictionnaire*, s.v., *flabellum*, *follis*, *uter*. The basic facts of history pertaining to these engines are still shown most completely by T. Ewbank, *A Descriptive and Historical Account*

paratus was known from the prehistoric
bellows (shown by Taccola, in either
heart-shaped or square form for blowing
air or pumping water, actuated by treading or by waterpower and cam action,
etc., Pls. 5, 7, all of which later became
part of such books as Biringuccio's
Pirotechnia, folios 109 ff). The cylinder
and piston structure was invented by
Ctesibius and described in classic terms
by Vitruvius (X, vii).

Taccola shows no familiarity with the
Vitruvian specifications, for instance
that such a pump should be of bronze,
that valves should be provided in the
cylinders and in a vessel above them, or
that the pistons should be rubbed with
oil. However it is clear from the drawings in Books I and II that piston action
is known to Taccola. He shows this
action also in elementary form, as used
in a valveless syringe or *schizzatoia,
ischitatorium,* S 97r.[73]

Simple forms of these various devices
may have been used to some extent
during the Middle Ages, although we
do not find them in manuscripts by
Fontana, Kyeser, the Hussite Anonymus,
Guido, or Villard. By contrast, a number

of improvements over the basic device—
or in some cases attempted improvements without real merit—are shown in
Taccola's work. Among them is a construction that foreshadows the more
recent leather piston and the modern
resilient piston ring: a piston in the
form of a sand-filled bag or *saccus* (II
85r, S 105v; *De machinis,* Lat. 28800,
27r). The present folio seems to emphasize a single point: the use of cylinders—
obviously of wood—which are "glued
and strongly tied" to make sure that
water will not "spray from cracks" in
the cylinder walls during the pumping
operation. Such spraying would not occur below the water level at the foot of
the pump or in a suction cylinder ex-
posed to air; apparently it had been
observed on the top or discharge side
of pumps during the upstroke.

It has been mentioned by Shapiro that
the valving shown at II 82v (Pl. 15),
which Taccola mentions here, allowed
the use of suction in the pump cylinder.[74] Taccola, on occasion, as for example in *De machinis,* 25v, expressly
states that the pump needs suction valving (*habere debet animellam sugentem
aquam*). There is no indication that he
considered such valving as new. Elsewhere (I 60v) he shows an automatic
valve system for a piston pump in a
sketch that suggests the presence of a
new idea rapidly jotted down. However,
it shows a valve system actuated by
linkage and ropes, not the flap valves
of a suction pump. Our impression
from Taccola's comment and the ancient

---

of Hydraulic and Other Machines, 16th ed.,
New York, 1870, pp. 205–270. However, he
knows about Taccola only from the secondary
copies in sixteenth-century editions of Valturio
and Vegetius, and his presentation is confused.
[73] The syringe has remained uncopied and
unpublished. The pump drawings are copied
frequently. For example, Taccola's I 58v (force
pump of square cross section, manually actuated by means of a weighted lever pivoted to
the top) occurs in Urb. Lat. 1757, 10r; Beck,
Beiträge, Fig. 347, and elsewhere. His check
valve constructions (I 72v, II 82v) have been
published by Berthelot, 1891, Fig. 46, and by
White, *Medieval Technology,* Pl. 7.

[74] S. Shapiro, "The Origin of the Suction
Pump," *Technology and Culture,* V, 1964, pp.
571 ff.

literature about pumps is that suction as well as force pumps and their valving, shown in II 82, were old and uninteresting in Taccola's time but that he was interested in smaller details, perhaps including the crank drive and surely including methods of tightening the cylinders.

It may be noted in this connection that Taccola's pumps do not have pipe connections. In this they differ from Roman pumps and bellows and even from Taccola's own bellows pump, I 31v (Pl. 7). His piston pumps usually extend vertically from a supply pool to an upper level where they discharge through a spout, as here; sometimes also by a directed jet, if the picture at III 44v–45r (Pl. 53) is taken at face value. The vertical cylinder merely functions as guide and container for a kind of reciprocating bucket apparatus. The probable reason for the complete lack of discharge piping connected with pumps is that the pumps were unable to operate under any significant discharge pressure; they immediately began to "spray from cracks." It appears that pumps in their late medieval stage of development, illustrated here, had only minimal application and water-lifting power. The ancient device was in need of redevelopment.

Perhaps a word should be said about the crank device shown here and elsewhere (II 82v, Pl. 15). Taccola uses crank devices also for mills, as shown for instance at IV 65r, and for bucket wells, as shown in *De machinis*.[75] On the other hand he also actuates piston pumps by levers or cams (for instance III 33v–34r,

44v–45r, Pls. 34, 53). When he uses cranks, he connects them to the load by ropes or rigid rods. He does not indicate whether these alternatives were new or conventional, but it seems possible from his presentation that they were known in fairly wide circles. His explicit notes on drive means always relate to structures other than cranks, for instance the simple winch drum handle of III 45v. Admittedly he draws the crank, on this page and at II 82v, in somewhat confused form, but this fact does not necessarily show that a new structure was presented to him. It may also mean that he simply did not stop to work out the repetitive, mechanical details of the device in their exact form. A copyist who had some technical understanding was neither misled nor impelled to claim that he had corrected anything.[76]

III 30r. Chain Pump. (Plate 27). Text in 10 lines and *taccola* vignette (Plate 28) on verso.

Of the various pumps that Taccola shows, he designates only the chain pump as something new in its entirety. He indicates (II 80v) that he learned of it through Bartolommeo Pasquini, otherwise unknown, and on the present folio he calls it a Tartar pump. Some of the Tartars lived in and throughout Eastern Europe, 1240–1480; others in Taccola's time lived in various coastal areas known to Italian merchants. As enemies of the Turks, these people were generally courted by European rulers and mainly by Sigismund.

<hr>

[75] Lat. 28800, 5v, 23v, copied, for instance, in Spencer Ms 136, 38v, 91v; Lat. 7239, 21v, 40v.

[76] This is clear from Add. Ms 34113, 23r. On the other hand the error is copied slavishly in Palat. 767, p 32, and even in Urb. Lat. 1757, 8v.

According to former accounts the chain pump came from China (not from a more western Tartar region) to Europe, through such travelers as Marco Polo or various missionaries; the accounts are speculative.[77] The account by Taccola suggests a transmission from China or elsewhere by Mongols and final stages of transmission (perhaps from Venice to Siena) by Europeans. It is one of the extremely rare cases where a link in early transmission of an Oriental technique is identified by a documented name, not a speculative attribution.[78] From Taccola it came, through his copyists, into general European practice where it was a distinct success, as is shown by Agricola, 1536.[79] Unlike other pumps, it did not depend on a valve system, and it was not limited to use in wide wells or low ditches, as is true of bucket chains, bucket wheels, and Archimedean spiral buckets.

So long as the machine was very new to the Western world, artisans of course had difficulty in building it and even in describing it. Taccola, while drawing it clearly, is groping for words to designate the principal series of elements on which it relies. He compares them with either mushrooms or spinning tops, while the literature of later practitioners sometimes speaks of "rags." Taccola makes it clear that they must be "neither tight nor loose" in their pipe or cylinder, as is also true of any piston. He does not show very suitable materials and does not specify tolerances; in fact, the concept of tolerances is unknown to him. He implies that builders should look to their own experimentation and ingenuity for the purpose of building this pump more easily and more successfully. The required ingenuity was in due course given either to Taccola himself or to someone whose work he knew, as is shown by his "sack pump." Yet it probably took further decades before the chain pump was developed to the extent shown by the later writers, in whose time it revolutionized the mining industry.

│ III 31r. Bucket Well with Treadwheel Drive. (Plate 29). Text in 10 lines and dragon vignette (Plate 30) on verso.

The treadwheel, known from biblical and classic sources and used in ancient times to actuate all kinds of winches for wells and building yards, as well as mills, is shown here as applied to a well. In the text the concept of a universal machine is mentioned, and beginnings are made toward recognition and formulation of things whereby such a machine can be "made to tick." Taccola formulates, at least in part, the reason why a larger treadwheel is preferable to a smaller one, the smallest possible being just above twice the height of a man. Mechanics no doubt had observed an advantage of larger treadwheels and had in their own groping terms attempted to describe it and explain it to Taccola.

---

[77] Ewbank, *Descriptive Account*, pp. 148–158.
[78] Suggested by A. Keller in a private letter. As to early reference to Tartars, see *Bellifortis*, X, fo. 130v for the comment that the battle scythe is used by Turks and Tartars. Also see: L. White, "Tibet, India and Malaysia as Sources of Western Technology," *American Historic Review*, LXV, 1960, p. 515.
[79] Palat. 767; p. 33, text p. 237; Add. Ms 34113, 23v; *De re metallica*, Book VI, pp. 188–196 (on the "six kinds of rag and chain pumps") in Hoover's translation, London, 1912, New York, 1950.

In modern terms we speak of a better ratio of moments of load and bearing friction; in Taccola's time these analytic terms and concepts were unknown. As is well known to historians of science, success in the effort that ultimately developed them was dependent on the rejection of Scholastic-Aristotelean nomenclatures, an achievement that became a reality in Galileo's time, two centuries after Taccola.

While not having the benefit of concepts and terms such as friction, load, moment, or even ratio, Taccola has an engineer's intuition of what tangible parts are decisive: wheel diameters and bearing constructions. Both of them no doubt had been studied by mechanics for a long time. Here their discussion enters into written form in one of the major channels of Western tradition. The formulation is based on consideration of various types of wheels and bearings, as may be noted from the classified list of mechanical elements at the end of this publication. It will be for others to determine whether or to what extent this channel of tradition led to further technical analysis or discovery.[80]

The treadwheel itself of course remained one of the most important prime movers for a long time after Taccola, together with other prime movers then known—animal drives, water drives, wind drives, and the strange clock-like mechanical weight drive. The treadwheel was superior to them with respect to starting, stopping, and reversing, as well as adjustment of operating speed. Brakes and clutches now used for these purposes were then only in process of development (Lat. 197, Hussite Anonymus, 3v).

III 32r. Siphon. (Plate 31). $6\frac{1}{3}$ lines of text, without vignette, on verso (Plate 32).[81]

For an understanding of Taccola's siphons it is necessary to consider his relation to the First Book of the *Pneumatica* by Heron, the Alexandrine writer who was active ca. A.D. 62. The book is also known as *De voto*, because its introduction deals with the void and with observations proving its occurrence in spite of Aristotelean fallacies. Parts I to VI of the Graeco-Roman work dealt with siphons: how to start them by suction (Taccola illustrates it at I 68r, Pl. 11); that they do not operate because of greater or lesser weight of water in their rising and falling tubes but because the water seeks equilibrium; that they can be built in various forms and with various controls at intake and outlet valves shown at I 73v (Pl. 12); and that mechanized starting is possible (as illustrated at I 60r, 62v, and 63r of Taccola's book). Heron, in parts IX to XLII of his book, after a short section on the liquid-lifting clepsydra (Taccola's I 75r), described a great many combinations of these devices with pipes, valves, and pumps, adapting them largely to pur-

---

[80] For the construction of late ancient machine bearings, see Beck, *Beiträge*, pp. 27–36. Taccola's drawing is reproduced in Palat. 767, p. 34, text p. 240, and Add. Ms 34113, 24r. The presentation is similar in the *Trattato* of Francesco di Giorgio. Only Leonardo began to analyze some of the machine elements to an extent greater than Taccola had done it.

[81] Add. Ms 34113, 24v, has both drawing and text, while Palat. 767 gives the text alone, p. 241. About early siphon developments, see Daremberg-Saglio, *Dictionnaire*, s.v. *siphon*.

poses of entertainment in gardens, grottoes, and pleasure resorts.

These entertainments became very popular wherever any trace of them was known, for instance at Byzantine and Arabic courts.[82] They also became extremely popular in the Renaissance. Several of them are shown in the copybooks based on Taccola's work, for instance Palat. 767, 1 to 4, and Add. 34113, 165 to 175 (datable ca. 1470–1500). In due course they found their way into early printed literature. It is clear that Taccola had limited and yet significant knowledge of Heron's work, but it remains an open question how he obtained it. The relatively pure manuscript copies of the *Pneumatica,* produced shortly before his time, were not translated for him, or he would have avoided some errors.

Taccola states that the descending pipe of a siphon should be one-fourth longer than the ascending one. Elsewhere (I 73v) he makes it one-third longer, and he does not seem to be aware of a discrepancy. He must mean that it should be substantially longer; more precisely, it should extend to a substantially lower level. Demonstrations of the type that he illustrates (I 68r) must have made this requirement apparent. Perhaps we should conclude that his "measures" must not be taken at face value and that they express qualitative rather than quantitative relationships. At III 28v, where he says the downpipe in a fountain

supply should be three times wider than the riser orifice, perhaps he only wishes to indicate that it should be substantially wider. It may be noted that when he gives numerical values he often gives widely differing alternatives. He names 1,000 steps and 500 steps as bases for the gradient of an aqueduct, III 33r; 6 feet and 8 feet for the thickness of a foundation slab, III 42v; and 16, 20, or 24 paddles for a waterwheel, III 45v. Obviously he will also accept intermediate values, and probably he will accept modified values. Quantitative rules more definite than these were hardly possible at the time. Only the most elementary units, denoting length, weight, and time, were current. No one had a concept of mechanical force (mass lifted times height of lift over square of time) or of more complex quantities such as energy or power. Antiquity did not develop such concepts, and the Middle Ages surely did not supply them.

Although such concepts were undiscovered, the quest for clarified knowledge was strong again, as it had been in Antiquity. Such demonstrations as Mariano records at I 68r (Pl. 11) were undertaken by fundamentally curious people. Their questions were primitive and their formulations vague, but they were determined to interrogate nature, not only books.

They also were inspired by a quest for power. In addition to the present siphon over a river and the tunnel under a mountain shown elsewhere, they wanted siphons to straddle mountains, transporting water from its source to the point of use in the most direct lines. This kind of proposal was known in

[82] There are many modern reviews of this old literature and also various medieval intermediaries. See G. Sarton, *Introduction,* I, 630 f, II, 632 f, about the books of the Banū Mūsā and of al-Jazarī, dating in the ninth and twelfth centuries.

various countries. Kyeser has it in his *Bellifortis*.[83] No doubt he, like the Hussite author and others, derived his information from medieval traditions based on ancient studies, just as Mariano did. They hoped that water in a siphon would ascend over, and perhaps onto, a mountain (I 105r).

The siphons remained inoperable when extended to a height that exceeded that of a very minor hill; 34 feet was and is the absolute limit. The phenomena were mysterious; therefore they continued to fascinate the people.[84] It took "only" two centuries after Mariano's time, that is, until Torricelli's time, until the gravitational pressure of the atmosphere was realized. It explained the phenomena of the siphon and precipitated a scientific and technical revolution.

| III 32v–33r. Aqueduct and Reservoirs. (Plates 32, 33). 10½ lines of text on 33r. No vignette, unless the fish shown in the well is intended as such.[85]

Taccola connects an upper masonry reservoir to a lower reservoir by means of an elevated aqueduct. He lets the masonry cure for two years, as he anticipates that the lime-cemented wall structure would soon become porous and leaky if it were not consolidated soundly. (By contrast, he specifies only ten days for curing a mere foundation slab, III 42v). As supports for the aqueduct, he shows round arches or "bridges"; he

seems to assume that it is known how to build them.

As to the slope, gradient, or *dependentia*, an analysis of which occurs at IV 58v, he specifies a descent of 2 feet for a distance of either 5,000 or 2,500 feet. Vitruvius (VIII, vi, 1) gives a single value, the reading of which is controversial; it is not the basis for Taccola's specifications. Nor does Taccola reflect the lesser gradients of Frontinus.[86] He probably uses typical slopes of Roman aqueducts surviving in his time. If his aqueduct has a slope of 2:5,000 and a 1 × 1 foot cross section, and if it runs full, it can supply about 1 cubic foot of water per second in the absence of loss by leakage. This amount was probably enough for the needs of a large hospital or medium-size city of the time. Comparable elevated aqueducts were proposed for Siena in the Trecento, but only underground aqueducts as shown at III 49 (Pl. 60) were actually used and extended.[87]

| III 33v–34r. Waterpower System for a Mill. (Plate 34). She-wolf emblem on 33v, with an inscription in lettered form. 5½ lines of text on 34r; pasturing sheep vignette on 34v.[88]

[83] Folios 59v, 61v, cited by Berthelot (1900), pp. 350, 353.

[84] C. de Waard, *L'expérience barométrique*, Thouars, 1936 credits A. Kircher, S. J., not Torricelli. In any case, the critical events occurred about 1610 to 1645.

[85] Copied in Palat. 767, p. 29; Ital. Z 86, 58v; Add. Ms 34113, 27v–28r.

[86] Frontinus, *De aquis*, 1899, pp. 162, 194, 195. About pre-Roman aqueducts, reservoirs, and irrigation, see Daremberg-Saglio, *Dictionnaire*, s.v. *aqueductus* and *fossa*; A. W. Van Buren in Pauly-Wissowa, *Real-encyclopaedie*, Supplement VIII, Stuttgart, 1956, *Bewässerungsanlagen*, cols. 9–16.

[87] Bargagli-Petrucci, *Le fonti*, I, p. 35 ff.

[88] Exact copies are in Urb. Lat. 1757, 101r; Add. Ms 34113, 28v–29r; Ital. Z 86, 58r (undoubtedly copied from Palat. 767, where this drawing, however, seems to be lost). As to the next chapter, III 34v–35r, only Add. Ms 34113 has a copy, 29v–30r, while Taccola's Sequel has a version, S 105v, which has been

This chapter may be considered the principal record of Taccola's "third invention," tested in 1427 and cited at I 61r. His purpose is to provide for a water-driven mill at the shore of a lake in the absence of natural "incline" or gradient. The present system provides for pump-actuated recirculation from the lake (here shown as a pond) into an elevated tank. A miller can thereby provide the needed fall of water to drive his mill. Near the lake he builds his elevated tank. Below this tank and above the lake he places his mill wheel and the mill driven by it. Variants of the system are shown elsewhere (III 45r).

What is the advantage gained by this indirect operation? The miller is not constantly tied to a tread-wheel or forced to use an expensive animal drive all the time that his mill operates. He can fill the elevated tank by pumping water from the lake, for instance in the morning, and devote himself to other chores during the day. The mill can nevertheless operate continuously if the elevated tank is large enough to let water run over the wheel and back into the lake for a whole day. On a small scale the system provides a mechanized and semi-automated mill. It is a man-powered equivalent of the tide mill shown in the next chapter. It may also be called a hydraulic equivalent of Taccola's weight motor system (I 15r, 57r; see Pl. 2 for one of these).[89]

Whether this system, in one or several of its forms, was new must be investigated elsewhere; little if anything has been published about it in the modern literature. Certainly it is the principal invention claimed by Taccola, other than those in the field of hydraulic construction (III 43). It appears that he promoted it by more than books and illustrations. He shows a model (I 74r) and mentions it later (III 57v); this model clearly served to demonstrate the operation of a mill by recirculating fluid, using mercury instead of water. This kind of drive was known in clocks. Did Taccola know this? Perhaps he wished to retain his knowledge of such details and for that reason kept the drawing of the model in his private notebook.

When Taccola addresses the Emperor on this page, it is on the plane of civic development in a very general sense. His emblem, the Roman she-wolf of Siena, recommends Siena and her works —presumably also her mill systems— to imperial consideration: *Senarum arma sum, civitatis virginis Marie, et sub imperii libertate filios meos lacto, quos semper devotissime imperatori sirenissimo ac invictissimo, domino domino meo, recomendo.* The words are lettered in the *libraria* hand, used by Mariano when he has God or a Saint speak to his sovereign: "I am the guard of Siena, city of the Virgin Mary. Under the Empire's

---

reproduced by Feldhaus (1910), p. 80 and (1954) p. 239.

[89] Folio 57 has an illustration similar to that reproduced on Pl. 2, and also has a fairly self-explanatory text: "Nota q*uod* qu*ando* vis sine aqua triticare si es in regione tufosa fac puteum cavernum p*ro* 200 brachios Et rotam

magnam p*er* altitudinem octo brachior*um* c*um* pondere facienti girari . . . . Et qu*ando* pondus rote tangit fundum puteii accipe simile pondus et applica ex a*parte* Et prius desciende p*er* scalas subterraneas p*rout* vides etc Et scolle a fune pondus Et fac ordinatim q*uod* quodlibet pondus conversur[u]s faciat rotam girare p*er* quartam horam. . . ."

freedom I suckle my sons, whom I recommend in everlasting devotion to my
lord, the most exalted and unvanquished
Emperor." He reserved technical details
for discussion with other masters, such
as Brunelleschi.

The mill system underwent change
later in Taccola's lifetime. In his later
treatise, *De machinis,* he replaced the
drawings described here in Book III
with very different ones. Still other
drawings, identifiable as a Mill Complex,
appear in all copybooks based on the
De ingeneis. In these versions recirculation by pumps continues to serve as one
of the major systems, but all structures
are shown in heavily modified forms,
both technically and graphically. Those
forms then became part of modern technical literature. The development is in
need of further study.

| III 34v–35r. Tide Mill. (Plate 35). Text
of 6½ lines and putto vignette on 35v.
(Plate 36).

Here and in a similar system, differ-
ently illustrated in the Sequel, S 105v,
Taccola shows a partial equivalent of the
preceding water systems for mills. In
this case the system is powered by natural forces that produce tides—forces
that were much marveled at, partly
"measured" in their dependence on
phases of the moon, and in no way
understood. They were reliable enough
to be used and also were large enough,
at least in the Adriatic although not in
the Ligurian Sea near Siena, to make
their use economically worthwhile. Taccola may have derived the information
for this chapter from the Adriatic region; as so often happens, the channels
of information are unknown. He also
knows that tides extend into rivers and
thereby into regions where mills can be
relatively accessible—a fact only partly
recorded before (for instance by Grosse-
teste in the early thirteenth century).[90]
It was apparently unknown to Galileo,
who tried with small success to describe
and explain the tides comprehensively.

| III 36r. Reversible Hoist (Brunelleschi's
Machine). (Plate 37). Also shown in
modified versions at III 37r, 38r. (Plates
39 and 41). The versos have a total of
34 lines of text. The vignettes are a rider,
a unicorn and a falcon (Plates 38, 40,
42), the latter being cited by Taccola in
a later note.

Folio 36 represents the famous cupola-
building machine devised and constructed by Brunelleschi in 1421, as can
be shown in detail by the record of the
Florentine *Opera del Duomo.* Here we
may compare Taccola's method of illustration and description of the machine
with an actual structure, particularly
since the machine is also shown in drawings by Giuliano da San Gallo, Leonardo
da Vinci, and others.[91] Taccola had the
advantage of knowing Brunelleschi
(S 107v). His description and illustration
are most likely to be based on Brunelleschi's own explanations. The language
used on 36v echoes Brunelleschi's famed
eloquence, rather than Taccola's laconic

[90] R. C. Dales, "The Text of Robert Grosse-
teste's 'Questio de fluxu et reflexu maris'
with an English translation," *Isis,* LVII, 1966,
pp. 455–474, especially p. 467. About medieval
tidemills, see L. Delisle in *Journal, British
Archaeological Association,* VI, 1851, pp. 406 ff;
A. Stowers in *Transactions, Newcomen Society,*
XXX, 1955–1957, p. 249.
[91] Scaglia (1955–1956) and Reti (1965) *passim.*
The documents are in C. Guasti, *La cupola di
S. Maria del Fiore,* Florence, 1857,

style, when it lists "three reasons" for building the machine and then is carried on to a fourth reason, restating the others in more general terms. This kind of enumeration occurs only once more in Taccola's work, where again Brunelleschi is quoted (S 108v).

The device shown here (fols. 36 to 38) serves to drive a hoisting mechanism, the classic description of which was given in the *Mechanica* of Heron (III 10). The basic hoist uses one end of a rope running over a pulley to raise a load. Another rope end running over a second pulley lowers a counterweight. A winch drum installed on a level below the pulleys actuates the rope. Taccola as well as Brunelleschi knew this basic type of machine (I 16v, 68v, sketches; SF 22v, 23r). They must have known it from field experience, since the textbook of Heron was forgotten and existed only in the form of untranslated and unknown Arabic manuscripts. The machine was, however, used in practice, for instance in bucket-well installations. There is evidence in Florentine documents of the Trecento and early Quattrocento that, a short time before Brunelleschi, such machines began to be used also in building yards. The *Mechanica* had been the classic text also for screw devices (III 19). As applied in the form of presses, they were used in practice, although the old textbook was lost. The other applications began to be tried.[92]

The Brunelleschian invention combines the old, counterweighted hoist with a new drive, using the old screw element for manipulating a gear unit. The combination was new at the time, even if all elements were old, and it was most successful and popularly acclaimed. The inventive structure is amply documented in the *Opera* books of Florence. In fact these documents also account for a number of minor parts, which Taccola disregards here in this drawing. Mariano, however, provides something more important than such details—the almost authentic explanation of that which Brunelleschi proposed and achieved. A description of the machine, probably based on popular account, was recorded by Vasari (1568). It has been clarified as pertinent documents and drawings were rediscovered. We now find that Taccola's text and illustration provide by far the clearest statement of the operation, in semiauthentic language reflecting Brunelleschi's ideas.

It describes what may now be called a motion-reversing clutch system. This system made it possible to let the working animal walk along its circular track, without reversal of its motion and attachment, and yet to reverse the rope drum so that it would at one time raise a load and at another time set down the load and return the rope. The reversals of the rope, which were needed frequently, could be effected speedily and without

---

[92] Beck, *Beiträge,* pp. 29–32. Screw and nut mechanisms are also shown as a means to lift an assault tower in Kyeser, *Bellifortis,* 33r (Berthelot, 1900, p. 351) and by Taccola, S 131v. In addition they were used as aiming mechanisms for cannons; see A. von Essenwein, *Quellen zur Geschichte der Feuerwaffen,* Leipzig, 1877, Pl. A IX (a German Codex antedating Kyeser) and *ibid.,* Pl. Z XIII (Kyeser), also Taccola, II 82r. Other applications of this mechanism, as well as the interest that they aroused, are described by Scaglia (1966), *passim.*

confusion by manual operation of the screw mechanism. So far as we know, this was one of the first applications of a clutch system. It must have contributed much to the sensationally rapid construction of the *cupolone.*

During its first few years the machine raised heavy timbers for the construction of a scaffold and crane unit, called *castello* and *stella,* on the walls of the tambour that were to support the cupola. When the *stella* was complete, the machine could deliver even heavier stone blocks, brick, lime, sand, and water for the herringbone masonry.

For more than ten years, the machine operated heavily and steadily as a hoist for a continuous supply of these materials. It appears to have taken an hour or two to raise a large load of a few thousand pounds, such as the major stone blocks, or as little as ten or twenty minutes for a smaller load of a few hundred pounds. During a typical day it must have raised some 10 to 15 tons to a height of some 200 to 250 feet and deposited each load at the proper spot. The operation was performed without major delay, breakdown, or reconstruction.[93] It was justly admired as a major achievement. Brunelleschi was honored for it by formal awards of 1421 and 1423, and he received total recognition as a master builder.

Taccola illustrates only the hoisting machine built on the ground, called *edifizio de' buoi* in the *Opera* documents. He omits the *stella* high above it, as well as other details, which are however shown in drawings by Leonardo and others. Among other things Taccola's drawings omit Brunelleschi's use of major, medium, and minor winch drums, a simple form of which he indicates elsewhere (I 19v).

Taccola also omits finer details of gearing. He of course is ignorant of modern methods of perspective. He shows, as in almost all of his pertinent drawings, gear shafts crossing one another in different planes, not shafts with mutually intersecting axes. With this kind of shafting, meshing cogs interengage inefficiently and with destructive friction as one contacts the other along spiral lines.[94] The arrangement is very inferior to another one in which the plane of one cogwheel is tangent on the other cogwheel. That one had been described by Vitruvius (X, v, 2) and had still been shown in the twelfth-century work, *Hortus deliciarum,* by the abbess Herrad of Landsperg.[95] In her work the good abbess shared Taccola's use of folded planes for shafts and wheels according to general medieval practice; but her showing of substance is much closer to the developed, ancient pattern of Vitruvius than is Taccola's drawing, which was composed two and a half centuries after her time.

Surely Brunelleschi did not construct such mutually crossing cogwheel shafts; with them his machine could never have performed as it did. We also have later and better drawings of his machine,

[93] As to some of these details Prager (1950) and Scaglia (1955–1956) had different accounts, which are now obsolete.

[94] A. G. Keller, "Mechanical Linkages. The Evolution of Engineering," *The Chartered Mechanical Engineer,* XIV, 1967, p. 323.
[95] See an edition of the manuscript edited by A. Straub and O. Keller, Strassburg, 1879–1899, fo. 112.

which show the shafting as described by Vitruvius and illustrated by the abbess Herrad.[96] Taccola may not have known the Vitruvian shafting, or may not have appreciated it, or may have been used to actual, inferior gear shaft systems, or may simply not have cared about it since he was more interested in the basic design of a clutch system. He had just learned of this system, and he concentrated on it as he taught this new part of engineering science. He clarified its representation by omitting or underemphasizing other elements, or at least we may assume that such was his purpose. Perhaps he also served his own interest as consultant in doing so. Giuliano da Sangallo, Leonardo, and Neroni, men who are much closer to a modern approach, saw the Brunelleschian machine as an object of historic interest rather than as a means for the systematic teaching of an engineering principle. They, as well as others, must have had access to the machine or its model two or three generations after its use. They sketched it and its appurtenances in substantially identical forms, more detailed than Taccola's drawings.[97]

They presented, so to speak, the old machine for the machine's sake, not a principle distilled from it, as Mariano does. In addition, of course, their drawings are superior to Mariano's with regard to the use of linear perspective.

The new principle may have been only of limited benefit to later generations of engineers. This machine is disregarded by the author of Urb. Lat. 1757, a copyist and developer who shows technical judgment and is probably Francesco di Giorgio. In later notebooks generally attributed to this author, screw devices are shown as means for purposes very different from those of the Brunelleschian clutch system. Other artist-draftsmen, who produced other copybooks, show that they no longer understand the Brunelleschian idea. They, or a reader of their unexplained drawing, "corrected" it to show the device of Francesco di Giorgio.[98] A hundred years later the work of another master builder, Domenico Fontana, became famous. His invention made frequent reversals of motions of the rope drums unnecessary, and the fame of his work may have contributed to the eclipse of the fame of the Brunelleschian clutch system, which was then forgotten except for popular stories recorded by Vasari. The constructors of St. Peter's cupola in Rome, Antonio da San Gallo the Younger and Michelan-

[96] Particularly the dimensioned and detailed drawings of Giuliano da Sangallo (S IV 8, 47v, 48r) and Bartolommeo Neroni (S IV 6, 49r, 49v). Also see Beck, *Beiträge*, pp. 19, 48 f. The crossed shafts occur also in Taccola's mill drawings (I 66v, reproduced by Feldhaus, 1910, p. 62, and in 1954, p. 240). Sometimes, however, Taccola shows an arrangement (I 57r, 59r, 66r; also see II 87r, published by Beck, *Beiträge*, p. 344), which is at least somewhat more similar to the Vitruvian gear shafting and the bearing arrangement needed for it.
[97] The drawings of Sangallo and Neroni cited in the preceding note show the gears of the Brunelleschian machine equipped with rota-

table gear teeth, *palei*, as does Add. Ms 34113, 87r. Leonardo shows them in a detail, *Cod. Atlantico*, fo. 391v–b, together with the perspective assembly view.
[98] Copied in Palat. 767, pp. 173, 174, 188, without text. Exact copies of Taccola's drawing occur in Add. Ms 34113, 26r, 26v (the translation of the text is erroneous); S IV 5, 84v, 93v, 97v.

gelo, could have used the clutch system again, but it is unknown whether they actually used it. A reversing clutch reappears in later times and in different applications.[99] We do not know whether its modern form is traceable to Brunelleschi and his illustrators or to new inventors.

Taccola's descriptions of the machine in three different forms appear on Plates 38, 40, 42.

| III 39r. Drive for Bucket Chain. 39v–40r. Winch for Drawing a Wagon up a Ramp. (Plates 43–45). Text totaling 14 lines. Vignette of hare-hunting dog on 39v (Plate 44); flying birds (intended as vignette?) on 40v.[100]

These pages illustrate devices of ancient or prehistoric origin, well known in the Middle Ages. Taccola seeks more or less universal rules, including rules for relative dimensions or speeds. He finds only beginnings of a formulation in this latter respect. For instance, the main wheel shown by him (39r) should have "thin pivots"; perhaps we should interpret this to mean, pivot pins thin enough to allow machining by whatever lathe is available. The wheel itself, by contrast, should be large and should have "teeth" (not shown), obviously to prevent the bucket chain from slipping. (A chain wheel driving a bucket drum, instead of being the drum itself as here, is shown by him at I 20v; both the wheel and drum are shown without teeth.)

As to the ramp system (39v–40r), a more complex variant appears at I 59v, where two wagons counterweigh one another, and each has a special unloading chute. Mariano chooses a simple version for insertion here, probably for didactic reasons.

| III 40v–41r. Quarrying and Handling Large Columns. (Plate 46). III 41v. Winch and Roller Frame (Plate 47) and notes on The Architect's Work. Text of 18½ lines and dove vignette on 41v (Plate 47).[101]

Further Brunelleschian developments appear on this page: methods for producing large columns, for transporting them, and for conducting the architect's work in general. The transporting method suggested here is Brunelleschi's patented invention. The patent itself said generically that Brunelleschi had invented "some machine or kind of ship by means of which he thinks he can . . . bring in any merchandise and load on . . . water for less money than usual," and the documents on the patented enterprise make it clear that the specific merchandise and load consisted of marble from mountains near the seacoast. Various winches, principally discussed and illustrated on these pages, along with a special flatcar, apparently

---

[99] See Giovanni Branca, *Le machine,* Rome, 1629, part I, fols. 20v, 21r, and 30v, 31r.

[100] Copied in Palat. 767, pp. 31, 98 (without text; p. 98 recopied in Ital. Z 86, 27v), and in Add. Ms 34113, 25r, 27r (with text). Also see Wiedemann-Hauser, "Heben von Wasser," pp. 126–128, 141–143, 151 f.

[101] Copied in Palat. 767, pp. 144, 145; Ital. Z 86, 49v–50r; Add. Ms 34113, 30v, 31r, 31v (with text, which Palat. 767 and Ital. Z 86 omit). In substance the illustration appears also in B. Lorini, *Le fortificationi,* Venice, 1609, p. 191. (This is the end of Book IV in Lorini's work. Also see his earlier edition, Venice, 1596–1597. The illustration on p. 191 includes a mechanism from Taccola's drawing at I 61r, and there are other reflections of *De ingeneis* in other parts of Lorini's famous book.) Documents of Brunelleschian origin, related to III 40v–41r, are reported by Prager (1946),

were built at the mooring place near the quarry. The winch, or a close variant of it, appears also at III 43v (Pl. 51) and also at a very prominent place, I 1r, the very beginning of *De ingeneis*, where Taccola may have inserted it in or about 1427 in front of *capitula* drawings that may date as early as 1419. The time of conception of Brunelleschi's invention is not recorded, but it was patented on June 19, 1421; it became controversial shortly before the end of 1425; and its use was undertaken between the end of 1426 and the end of 1427. As a commercial enterprise, it was a complete failure; technically its value was clear to Taccola who presents it as a major concept. He is likely to be correct; new concepts are established by trial and error.

The mood of his chapter reflects the controversial nature of the enterprise. In fact, Brunelleschi's own words may be echoed here in Taccola's handwriting. This is suggested by an unusually heavy incidence of Italianisms, such as *comodo* for *quomodo*, *cui* for *qui*, and *grositiei vel dicamus latitudinis*, along with the usual *cielte*, *aculeis*, and so forth. Somebody else's words rather than Mariano's are also heard when he designates a compass as *sextus*. In addition, the quarrying operation is described in terms that sound as if an earlier description was paraphrased by someone less familiar with it than the original writer. Pliny (XXXVI, 1 to 18) may be the original writer, though his name is not mentioned here. His discussion of architecture is remarkably similar to Taccola's. Both writers combine in one chapter three heterogeneous

topics: special ships for marble transport (Pliny, 1–3), marble quarrying methods (4–13), particularly the production of obelisks (14, 15), and methods and achievements of architects in general (16–18). The combination seems sufficiently odd to make it probable that Pliny, directly or indirectly, stands behind Taccola's chapter, although Taccola or some intermediate removed certain details, addenda, and irrelevancies that he found in the ancient encyclopedist. Among the omissions are Pliny's bemoaning of luxury and dissipation, his remarks on marble sawing and polishing techniques, and his excursus on uses of obelisks and pyramids as gnomons.

Vitruvius (I, i) had also discussed the architect's work. However, the echoes here, as elsewhere in Taccola's treatise, do not lead us to this master. They may lead us to combined influences of Pliny and Brunelleschi or even to the possibility that Taccola discussed Pliny with Brunelleschi. The personality of the irate and pragmatic Florentine may well be reflected here, although his name is not mentioned.

III 42r. Cofferdam Method of Laying Foundations. (Plate 48). Taccola resumes his use of the format, showing a 20-line text and a squirrel vignette on the verso, (Plate 49).[102]

The drawing shows, in the usual diagrammatic way, one of the frequently recurring objects of Taccola's work, the *turris que est in mare*. It shows this tower complete, with its foundation visible,

[102] III 42 is copied in Urb. Lat. 1757, 46v–47r; Palat. 767, p. 39 (text only), and Add. Ms 34113, 32r, 32v (drawing and text). The copyist in Add. Ms 34113 introduces minor textual additions.

while the text describes the foundation-
laying process.

It fails to mention one of the elements
of this process, the bailing out or pumping out of water after construction of a
cofferdam, which Mariano had, however,
mentioned (I 61r) in his note about four
inventions of December 1427. Other details are also omitted. Yet the chapter
demonstrates Mariano's power of technical description. It describes a process,
known to experts but complex for lay-
men of his time, in terms that may pass
even now as a model of clarity and conciseness. Of course, in reading it, we
must be prepared to accept him as a
man of his time, not of our time. His
explanation of the troubles caused by
saltwater is now obsolete, as are his
references to a "tower," a "circle," and
a submarine "mountain." However,
within the limitations imposed by time-
conditioned terms and concepts, he
outlines the process in almost classic
form. He alludes to the prerequisites
and then describes about a dozen successive stages of the process in as many
sentences, progressing lucidly, and by
no means mechanically, from step to
step.

One could wish that notes left by other
artist-engineers were as clear as these.
Certainly the words of Vitruvius, where
he tries to describe a substantially similar process (III iv, V xii, and VI viii),
are less pertinent and evoke a less comprehensive picture. More comparable in
form and quality were the terms of
Heron in Book III of his *Mechanics,* the
vastly improved equivalent of the Vitruvian Book X. In view of remarkable
parallels, to be noted also in Taccola's

next chapter, it appears that some Western medieval residue of Heron's book
must have been known, although modern scholars of this Alexandrine work
know nothing of it. Perhaps we have
here a late medieval renaissance of
Heron, preceding the modern renaissance of Vitruvius that so absorbed the
minds of Taccola's followers. However,
other writers of the late Middle Ages—
Villard, Guido, Kyeser, the Hussite, and
Fontana—know nothing of it. In modern civil engineering the structure of
III 42 is called Open Caisson with Bal-
last Pocket. Earlier illustrations are not
known at present. Mariano does not
claim it as his invention.

III 43r. Box Caisson Process. (Plate 50).
Winch and Roller Frame, on verso. (Plate
51). The verso has 16 lines of text and a
bear vignette (Plate 51.)

This drawing of a box caisson illustrates the invention of 1427 that Taccola
claims for himself in the field of foundation work (I 61r), and the text describes
it in terms closely related to the Brunelleschian column transport method (III 41).
The same special boats, flatcars, and
winches are used, and they are cross-
referenced in the text. Here they manipulate the foundation of a building.

Taccola starts again with a brief statement of the problem: the structure built
by the conventional method (III 42) is
all too easily destroyed by the forces of
nature. The solution is to build the structure in a sheltered location, on ships,
later to float it to its proposed site, as
shown in the drawing, and there to
sink it so that it will stand in the destructive seawater but not be harmed by
it. Related proposals had been made be-

fore but, for all that we know, had been forgotten. Taccola now presents a rather specific description, which apparently was harder to forget.[103] The principal engine used is the fourth invention of 1427, the box caisson. It is a device similar to a pontoon. Brunelleschi's patented boat was rather similar, and so were the pontoons and caissons in the *capitula,* which were produced to raise and lower harbor chains and Chevaux de Frise for the Sienese. Taccola himself states that the caisson wall can be of a well-known form commonly called palisade, *lo stechato.*

He then describes the manipulation of the wall or foundation by the caisson structure. Pairs of additional caissons or "ships" have been provided, each with its own aperture and valve for flooding it. These, if not the central caisson and its valve, were known in Heron's *Mechanica,* and they recur in the early medieval *Mappae clavicula,* but both descriptions shown in manuscripts known today were obscure.[104] A specific purpose and the exact procedure are stated

for the first time in Taccola's laconic text, coupled with his diagrammatic picture. Additional caissons are shown as rowboats. When the unit has arrived at the site, which no doubt has been dredged and flattened before, the valves are opened in all caissons and the foundation settles to the ground. Since the cement has previously been allowed to bind, seawater does not affect it.

It is hardly to be expected that techniques of such complexity and relative novelty, known only from practitioners' traditions and improvements, then improved further by Taccola himself, could be described at once with finality or perfect clarity. Indeed, perfect clarity is not to be found here, and Mariano, as mentioned, may not even aim at it. Removal of water from inside a cofferdam is a necessary step in the method described on folio 42, for the ramming down of clay; however, this operation is not mentioned. Folio 43 speaks of an element "d" but does not show it now, and we do not know whether it ever showed it. This latter folio, like folios 40v–41r and 44r, is challenging also in that it shows rowboats, while the text calls for caissons, *casse* or vases, *cinbe* arranged for controlled immersion and submersion. Perhaps ordinary rowboats were used for experiments and tests.

Some details relevant to the method shown here on folio 43, although not explicitly cross-referenced here, are to be found in Book I. They include the construction of a caisson *cum tabulis contiguis ac fictis* (I 44r, 44v) and the required dredging of the ground and measuring of its elevations (I 46v). Details are also shown in Book I for a more con-

---

[103] See Daremberg-Saglio, *Dictionnaire,* s.v. *pons;* Berthelot (1900), pp. 290, 336, 340 ff. Taccola's version is claimed by him as an invention, I 61r. It was copied precisely in Palat. 767, pp. 143, 240; Ital. Z 86, 49r; Add. Ms 34113, 33r, 33v.

[104] See A. G. Drachmann, *The Mechanical Technology of Greek and Roman Antiquity,* 1963, p. 109 on *Mechanica,* III, 11; T. Phillipps, in *Archaeologia,* XXXII, 1857, pp. 183, 209, on *Mappae clavicula,* ch. 102: V. Mortet, "Un formulaire du VIIIe siècle pour les fondations d'èdifices et de ponts," *Mélanges d'archéologie,* 1915, pp. 233 f, 246 f. Mortet's extracts from the *Mappae Clavicula* refer to the "triangulation of the height of an object when the foot is accesible," which may be related to Taccola's *mensura* drawing on III 27r.

ventional method as described on the preceding folio (III 42), including earth borers, cofferdam elements, and pile drivers (I 47r, 47v, 50r, 50v, 54v, 55r).

Taccola, and at least one of his impor tant followers, clearly recognized the importance of the techniques summarized in the present folios 42 and 43. In the Sequel of *De ingeneis* (S 108r, 109r, 110r) and also in the *De machinis* (Spencer Ms 136, 12r) there are clear and detailed accounts based on these folios and developing these methods. In Alberti's treatise *De re aedificatoria* (IV 6, on bridge piers), probably written during Mariano's lifetime, the reflection of one of these methods may be found, differing only in that some details and all illustrations are omitted and that neoclassic Latin takes the place of the medieval idiom. Then, about a decade after Taccola's death, comes the author of Urb. Lat. 1757, probably Francesco di Giorgio, who copies the first of the present folios twice (10v and 46v–47r) and the other folio once (47v–48r), leaving space for further discussion (48v–49r) but leaving it blank. The later *Trattato* of Francesco di Giorgio then gives a description in Italian of these procedures in more modern, technical language. The writer complains about the difficulty of combining text and illustration when treating of hidden things such as *il fondare in mare*, which he still calls one of the *nuove ed inusitate invenzioni*. The terms that Francesco uses here are an apparent reflection of Taccola's expressions, even if the information transmitted here should be also supported by other and possibly earlier sources, unknown to us.

Francesco then describes and illustrates the exact structure of folio 42, followed by that of folio 43. His description of the latter combines brief indications of Taccola's Book III with pertinent details from his Books I and II. Four pontoons, *barconi*, are constructed in such size as to match the length and width of the proposed edifice, so that when they are tied together head to head, they form a rectangle of corresponding clear size (as partly shown in S 108r, 109r, 110r, 112r, and somewhat more clearly in I 50v). On these pontoons a crane is erected as shown in a drawing at I 61r, which accompanied the note of 1427, and more clearly in the later treatise of Taccola, copied for instance in the Spencer Ms. 136, 12r. It serves to suspend a central caisson, *cassa*, and ultimately to let it down into the water; while the caisson is suspended, the foundation is constructed in it.

Francesco has nothing to say about technical details of the construction of the caisson, the joints between its wood beams, the shaping of its bottom, and the guide structures to ensure its placement at the exact spot designated for it, all of which Taccola had illustrated (I 44r, 44v, 47r, 50r). Francesco also forgets to mention other important details, for instance that valves are provided in the pontoons (III 13v) or in the caisson (I 61r) or in both. However, he mentions that the bottom of the pontoons or caisson can be made flexible to facilitate its ultimate adaptation to the ground underwater—another improvement taken from Taccola (II 81v). Undeniably, Francesco's eloquent text, paralleled also by the text of Alberti, is clearer to modern readers than Taccola's "veiled speech," although it adds little or nothing by way of substantive improvement or invention, at this point.

In due course the caisson method appears in actual use. Beginning in the seventeenth century applications of the method are mentioned in the literature relating to the construction of foundations for the piers of famous bridges in Paris, London, and other major cities.[105]

|III 44r. Recovery of Sunken Columns and Treasures. (Plate 52). Text on the same page, 44r. Either the diver shown on this page or the hog shown on the verso, Plate 53, or possibly both, may be meant as a vignette for this chapter.[106]

The operation depicted here is related with those of the preceding folios 41v–42r and 43r, and Taccola makes actual reference to the wagon and winch "shown in back" and to the "aforesaid vessels." The operation is also related to that of a modern dry dock. In a developed system of this type the engineer would generally use deep caissons rather than the flat rowboats shown here, but in Antiquity and the Renaissance, writers describing such systems referred to conventional boats. This is true of Pliny (XXXVI, 67 f), who describes a method used by ancient Egyptians for transporting obelisks. In Taccola's time, Brunelleschi may have used an operation of the type shown here when he lost and later retrieved marble blocks or columns in connection with the operation initiated in accordance with 41v–42r, which may also be called his patent operation. Then, shortly before Brunelleschi's death, similar salvage operations were undertaken in Lago di Nemi to raise the so-called Ship of Trajan. Alberti mentions this work (V 12, VI 6), citing Pliny. It may be considered one of the major treasure-retrieving operations of the Renaissance.

|III 44v–45r. Water Supply System for Mills, and Discussion of Different Waterwheels. (Plate 53). 16 lines of text and a donkey's head as a vignette on 45v (Plate 54).[107]

Taccola presents a variant of the technical contents of III 33v–34r with a text about different waterwheels. He indicates that the earlier Tuscan wheels were of impulse type and driven by swiftly running brooks, and it seems probable although not certain that he means vertical-shaft units, also known as "Greek" waterwheels. He also suggests that larger, overshot wheels, driven by slower rivers, were later introduced.

Again, he proposes recirculation of water by piston pumps, the arrangement claimed by him in 1427 (I 61r). For once, he calls for a pump that has metal pistons and cylinders, *cannae agentes et patientes de metallo*. He has nothing to say about the Vitruvian use of bronze but proposes, strangely, the use of tin, *stagno*. The drawing reappears in Urb. Lat. 1757, 101r, with a seemingly slight

---

[105] See D. B. Steinman and R. S. Watson, *Bridges and their Builders*, New York, 1957, p. 88, on the caissons used by F. Romain in founding piers of the Pont Royal in Paris. Also see H. Straub, *A History of Civil Engineering*, London, 1952, Cambridge, Mass., 1965, pp. 132, 151.

[106] III 44r is copied in Palat. 767, p. 146; Ital. Z 86, 50v; Add. Ms 34113, 34r.

[107] III 44v–45r, is copied in Palat. 767, p. 100; Ital. Z 86, 34r; Add. Ms 34113, 34v, 35r, 35v; S IV 1, 107v–108r. The literature about medieval mills is large. See White, *Medieval Technology*, pp. 80 ff.

but actually most ill-contrived modification: a pump-actuating camshaft driven by the pump-actuated waterwheel itself, *a moto continuo*, 98r. This *ignis fatuus* then flickers through the later literature, while Mariano's system is forgotten. Neroni has a confused drawing of the *moto continuo* version (Cod. S IV 6, 41r).

| III 46r, 47r. Builder's Cranes. (Plates 55, 57). A total of 12 lines of text on the versos, with vignettes showing, respectively, a dragon (Plate 56) and a bat (Plate 58).[108]

Machines of the general types shown here were probably used for subordinate purposes on the Florentine *Opera* building grounds and elsewhere, both on the ground and above on the walls, as was required from time to time. Taccola specifies the differences between the crane here and those in general use by noting its mobility and the placing of the winch on the platform at the top. It is not apparent why Taccola should have included the cranes among the achievements in a book expressly termed, "not (previously) used," unless it be an excess of consideration for preferences that Brunelleschi may have suggested.

| III 47v–48r. Mill in a Mountain Landscape. (Plates 58, 59). No text. The picture of a small dog on the verso (Plate 60) probably was once intended as a vignette for the present chapter.

At this point the interests of Taccola the artist prevail over those of Taccola the engineer. Not only is nothing shown

to indicate the operating mechanism for the conventional millstone, but the entire setting is utterly fantastic. The presence of the vignette indicates, even in the absence of text, that the picture belongs to the original parts of the manuscript and that it was originally intended to be described somehow. Mariano may have changed his purpose in the process of developing the chapter and may have left the drawing merely as an ornament. The millstone shown here appears to be driven by the vertical-shaft waterwheel mechanisms suggested at III 33v–34r and proliferated in the Mill Complex.[109]

| III 48v–49r. Water Main and Outlet. (Plate 60). 9½ lines of text and a rooster vignette (Plate 61) on 49v.[110]

In this drawing Taccola shows the great pride of his home town, her *butini*. As usual, the drawing is very diagrammatic, and the description is limited to matters that at the time appeared essential for basic instruction. Taccola shows specific and more realistic structures of these underground aqueducts in the Sequel of his work and in his later treatise. As shown by some of the later disclosures, the water was (sometimes) conducted through pipes interconnected by spigot joints and cemented into the bottom slab of a covered walkway. The later drawings also show variants of the "high tower" outlet (S 112r, 113v–114r).

[108] The cranes, III 46r, 47r, are copied in Palat. 767, p. 186; Ital. Z 86, 34v; Add. Ms 34113, 36r, 36v; S IV 5, 87r, 97v. All of these devices are well known in history, as shown by Beck, *Beiträge*, pp. 37–57.

[109] See Palat. 767, pp. 60f, 63ff, 69, 74f, 78, 83, 87. The present drawing, III 48r, is copied only in Add. Ms 34113, 37r.
[110] Exact copies: Palat. 767, p. 29; Ital. Z 86, 58v; Add. Ms 34113, 37v–38r. Certain wall dimensions, as given here by Taccola, were usual in Sienese *butino* practice (Bargagli-Petrucci, *Le fonti*, I, p. 41)

Fundamentally we have here a U-tube system, as in III 28 (Pl. 23), but with larger pipes. Taccola shows recognition of the pressure problems encountered in such a system when he says *oportet quod ipsa tunba sit valde constructa*. The ways of how to solve the problem were for others than his imperial reader and perhaps also for others than Taccola himself.

| III 50r, 51r. Cranes. 52r, 53r. Ladders. 53v–54r. Amphibious Vehicle. (Plates 62, 64, 66, 68, 69, 70). The texts on versos total 37 lines, at Plates 63, 65, 67, 69, and 71. Owl and crow vignettes for the cranes, monkey, and dead hare for the ladders, and "small and large fish" for the vehicle.[111]

In these drawings of cranes and mobile ladders, most elementary mechanisms are shown in crude form. Vitruvius would have agreed with later builders that the illustrations are obvious and uselessly repeated. They seem to be due to haste in trying to complete the book. Proposals for mechanized vehicles are ubiquitous in the old books.[112] Taccola may include the amphibious vehicle only as a joke; his statement about amphibious oxen seems very odd. The amphibious wagon does seem to show imagination, especially in the idea of using pinions as wheels to prevent sinking or slipping in the mud.

| III 56r. Floating Mill. (Plate 72). Vignette of a snake on the verso (Plate 73).[113] No text. There is no folio 55.

The floating wagon of the preceding chapter is here followed by a floating mill. Constructions of this type were well known and widely used at the time. The Byzantine historian Procopius, ca. A.D. 550, attributed their introduction to one of his contemporaries, as has been noted for instance in Singer (1956). Taccola evidently shows their undershot arrangement as an addition to the more basic types that he had discussed at III 45v. For some reason he then shows the floating arrangement once more at IV 66r.

| III 56v–57r. Siphon as Water Supply for a Mill. (Plate 73). The verso of folio 57 has $10\frac{1}{2}$ lines of text and a leopard vignette (Plate 74).[114]

[111] Copied in Palat. 767, pp. 99, 109; Ital. Z 86, 75r, 57r; Add. Ms 34113, 38v–39r, 39v–40r, 40v–41r. Ancient machines had been more developed, according to the fairly reliable reconstructions of Vitruvian cranes (X, ii, 1–10) and gear drives (X, v, 2), as shown by Beck, *Beiträge*, p. 42 f, and Drachmann, *Mechanical Technology*, pp. 143–146. Relatively developed machines are also shown in the Hussite manuscript (see Berthelot, 1891, pp. 440 ff; Beck, *Beiträge*, pp. 271–273).

[112] Berthelot (1900), p. 313, traces the medieval illustrations of chariots "driven from inside" to such ancient prototypes as the *cattus* shown on the arch of Septimius Severus.

[113] Copied in Palat. 767, p. 101; Ital. Z 86, 77r; Add. Ms 34113, 45v. Other floating mills in Palat. 767 appear on pp. 82, 85. Stationary mills with waterwheel mounted on a horizontal shaft, with bottom impact: *ibid.*, pp. 59, 77, 79; top impact: *ibid.*, pp. 62, 71, 73, 86. They add proposed refinements, mostly of useless nature, to the simple mills shown by Taccola. The strange complex of drawings appears also in the autographs of Francesco di Giorgio and the copybooks based thereon, including those edited by Maltese (1967), I, Pls. 62–75; II, Pls. 325–331.

[114] The drawing of a technically similar siphon is in Kyeser, *Bellifortis*, 61v. Taccola's drawing is copied in Palat. 767, p. 111; Ital. Z 86, 56r; Add. Ms 34113, 41v–42r; S IV 1, 108v, and in various illustrations in the *Trattato* of Francesco di Giorgio and its copies; see Maltese (1967), I, Pls. 66, 70, 75, also 78, 81. About later printed copies, see Reti (1963), *passim*.

The transmission of recorded knowledge invariably includes admixtures of transmitted error. Someone, it seems, had proposed that a duct could bring water over a mountain, perhaps to avoid the cost of tunnels. Those who copied it, with or without improved mills, accepted the idea as a most attractive one. They copied and recopied it, well into the seventeenth century. When such siphons failed to work, builders probably were blamed for faulty workmanship. In the long run, small-scale and medium-scale experiments may have led to more fundamental doubts and to the recognition of an absolute limit for the height of an operable siphon. This recognition was reached some time before Galileo's time. Then, finally, came the elegant experiments that brought the explanation.[115] Until then, conceivably something was gained even by the transmission of erroneous knowledge.

[115] That is, the experiments of Torricelli-Viviani (1643), leading to those of Pascal-Périer (1646–1647), which put an end to scholasticism.

20. First portrait of Sigismund **(Parchment cover)**.

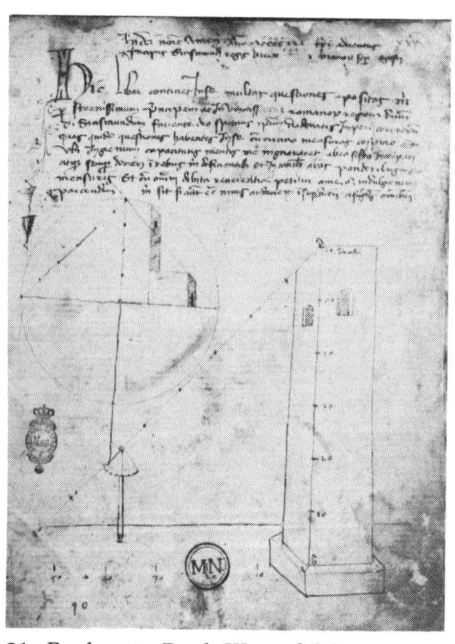

21. Preface to Book III, and *Mensura* **(III, 27r)**.

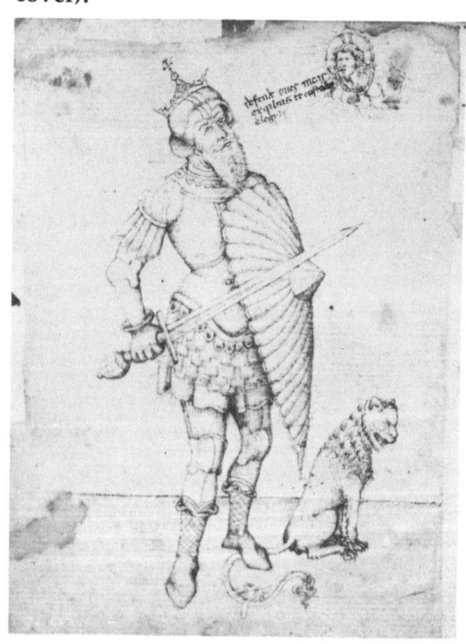

22. Second portrait of Sigismund **(III, 27v)**.

Fol. **27r** (Plate 21)

*In dei nomine amen anno MCCCCXXX
tempore adventus principis Sicismundi
regis buen* r[o]manorum se[m]per agusti*

*Hic liber continet in se multas questiones
apositas mihi per stren[u]issimum princi-
pem ac invinctissimum romanorum regum
dominum dominum Sicismundum favente
deo spe[c]tans ipsum diadematis imperii
coroneatum quas quidem questiones habentes in se cum circino mensuras cospicio / Et
ubi ingienium capacitatis mentis mee ing-
noraret / ab eo sisto increpari atque semper
docieri in rebus mihi deficientibus Et in
omnibus aliis ponderibus Et mensuris / Et
cum ommni debita reverentia petitur a me
M. indulgentia quod parcendim mihi sit
si autem esse[m] nimis auduciem in super-
dictis et singulis omnibus.*

* Some letters are lost here.

In the name of God, amen. 143[1], at the
time of arrival of Sigismund, the ruler,
King of Bo[hemia and] Rome, of permanent majesty.

This book contains [answers to] many
questions, posed to me by the most
mighty ruler and most unconquered King
of Rome, the lord, lord Sigismund, when
about to be crowned by the grace of God
with the diadem of the Empire. The
questions, I think, have [answers through]
measurement by compasses. If, at any
point, the intellect [or] capacity of my
mind be found insufficient, I stand ready
to be admonished by him, and always
to be taught of the things that have
escaped me and of all the various weights
and measurements. If I, M[ariano], be
too daring in any or all of the aforesaid
things, then, with all due reverence, I
pray for indulgence and that I may be
forgiven.

23. Fountain. *Incipit* of Book III **(III, 28r).**

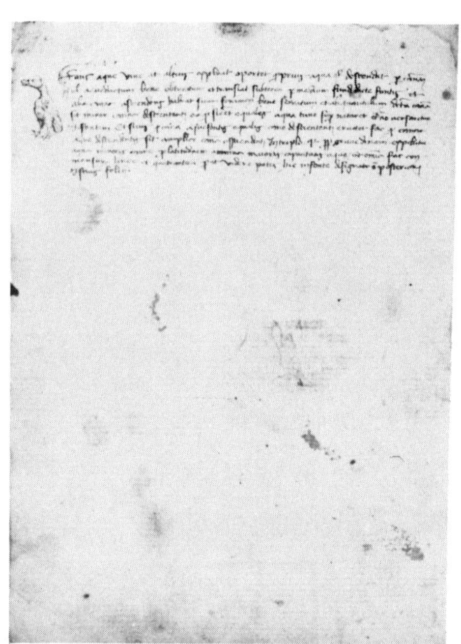

24. Text relating to the fountain **(III, 28v).**

Fol. **28r** (Plate 23)

*Incipit Liber Tertius de ingeneis ac edifitiis
non usitatis.*

*Fons aque vive ut altius expluat oportet
quod prius aqua e[i]us descendat per cannam vel aqueductum bene obturatum et
transiat sub terra per medium fundi illcle
fontis et alia canna asciendens habeat suum
foramen bene serratum et aliquantulum
dicta canna sit minor canna descententi eo
quod si essent equales aqua tunc super-
nataret et non versaretur in fontem Et si
vis quod canna aque fontis equalis can[n]e
desciententi eructet fac quod canna aque
descendentis sit amplior canna assciendenti
in triplo / quia propter gravedinem ex-
pelulur aqua minoris canne per latitudinem
a canna maioris capacitatis aque / Et
omnia fac con mensure lineè et quatrante.
proul videre potes hic in fonte designato
in posteriori istius foleii.*

To let the water jet spring forth to some
height in a fountain, the water should
first descend through a well-sealed pipe
or duct, and pass below the ground
through the middle of the bottom of said
fountain. Another pipe, ascending,
should have its aperture well connected
thereto, and this pipe should be somewhat shorter than the descending pipe,
because if they were equal, the water
would [only] overflow and not be thrown
into the fountain. If the pipe in the
fountain shall discharge in amount equal
to [that in] the descending pipe, arrange
that the descending water pipe be three
times wider than the ascending pipe,
because then the water of the pipe that
is smaller in width is expelled due to
gravity, by means of the pipe having the
greater content of water. Make everything with measurements, lines, and
quadrant, as you can see it here in the
fountain design on the other side of
this sheet.

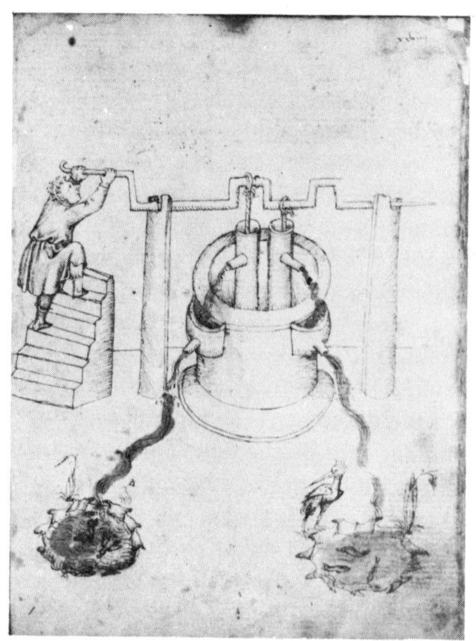

25. Piston pump **(III, 29r)**.

26. Text relating to the piston pump **(III, 29v)**.

Fol. **29v** (Plate 26)

Puteus aut cisterna prout videtur in pos-
teriori figurata / potest aqua attingi cum
cannis tangientibus intus aquam et ordinari
quod dicte canne sint de lignamine collates
atque ligate stricte ut aqua ne versetur
per earum rimulas. Et intus edificietur
ingienium duarum animellarum positarum
per contrarium unius alteri opposite prout
patet in designo in libro primo leonis a
fo[lio] 49 Et clarius apparet in secundo
libro draconis fo[lio] 79 designato ex manu
prop[r]ia Ser Mariani Jacobi decti Tachola

A well or cistern, as seen in the picture
on the other side, can be supplied with
water through cylinders extending into
the water. Let it be arranged that said
cylinders be of wood, glued, and strongly
tied so that water may not spray from
cracks therein. For its inside let there
be built an apparatus of two flap valves,
one placed counter and opposite to the
other, as appears from the drawing in
the First Book, of the Lion, folio 49
[I 53r], and appears more clearly in the
Second Book, of the Dragon, folio 79
[II 82v], drawn with his own hand by
Ser Marianus Jacobi, called Taccola.
[See Plate 15.]

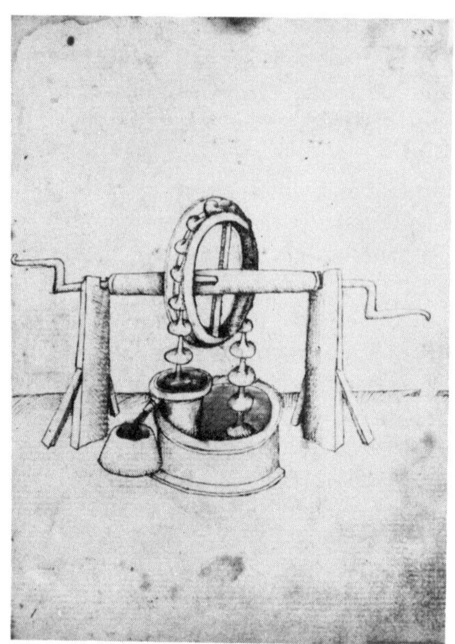

27. Chain pump **(III, 30r).**

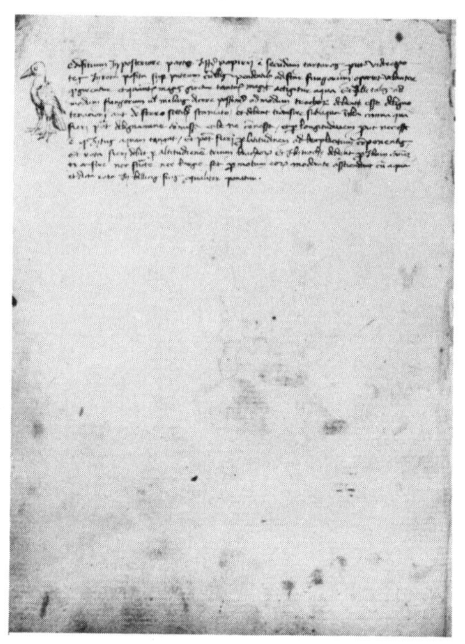

28. Text relating to the chain pump **(III, 30v).**

Fol. **30v** (Plate 28)

Edifitium in posteriore parte istius papirii
est secundum Tartaros prout videre potes
in rota posita super puteum cum illis
pendentibus ad istar fungorum / oportet
velociter quod girentur / et quantum magis
giretur tantum magis actigitur aqua Et
ille talos ad modum fungorum vel melius
dicere possimus ad modum trochorum
debent esse de ligno tenaciori aut de ferro*
sottilii staniato / et debent transire sub
aqua in illa canna quam fieri potest de
lignamine conmissa et bene connessa / et
per longitudinem prout neciesse est quod
intus aquam tangat / Et potest fieri per
lutiludinem ad bcn placitum conponentis /
Et rota fieri debet per altitudienem trium
brachiorum Et illi trochii debent per illam
cannam transire nec stricte nec largie set
per motum eorum moderate assciendant
cum aqua. et dicta rota in billicis suis
equaliter ponatur.

* The text shows *debente,* with a line drawn
through the final *e.*

The device on the other side of this
paper is in the manner of the Tartars.
As you can see, [it has] a wheel placed
over the well, with hanging attachments
similar to mushrooms. It must be driven
around rapidly. The more [rapidly] it is
driven, the more water is obtained. The
links here, similar to mushrooms, or as
we may better say, similar to spinning
tops, must be made of rather strong
wood, or of thin and tinned iron. They
must pass, under water, through the
pipe, which can be made of jointed and
well-connected wood, and of such length
as is necessary that it extend into the
water. As to width, it can be made as
the builder desires. The wheel should
be made six feet high. The spinning tops
must pass through this pipe and be
neither tight nor loose but must, throughout their motion, readily rise with the
water. The wheel must be placed evenly
in its bearings.

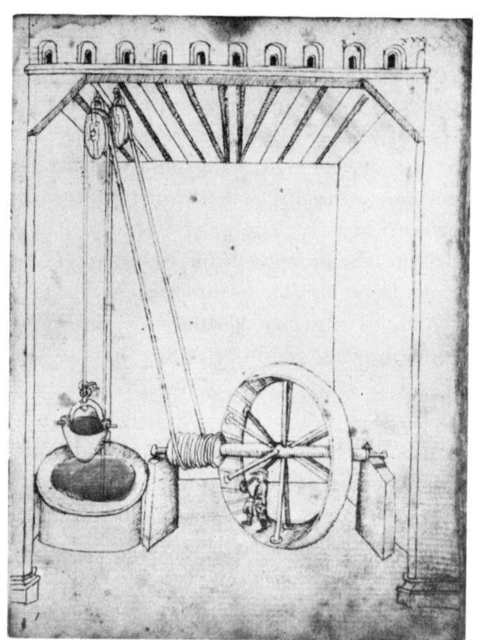

29. Bucket well **(III, 31r)**.

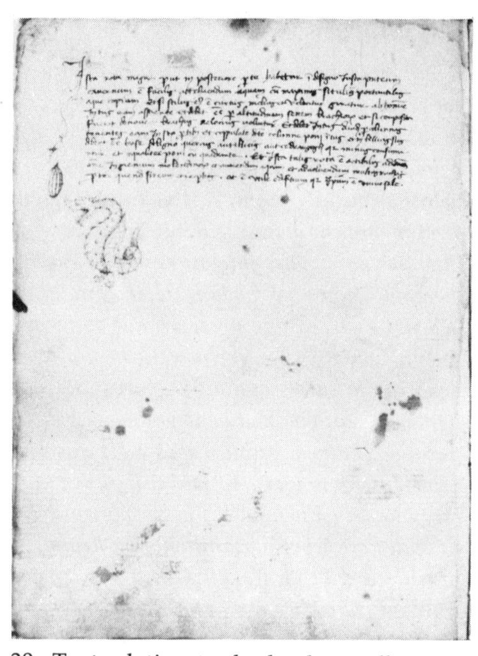

30. Text relating to the bucket well **(III, 31v)**.

Fol. **31v** (Plate 30)

Ista rota magna prout in posteriore parte
habetur in designo iusta puteum cavernum
est facilis attrhaendum aquam cum maxi-
mas situlis portantibus aque copiam Et si
stilus e[i]us est curtus mellus et velocitur
giratur ab homine intus eam assciente Et
debet esse per altitudinem settem brach-
*iorum et si conposita fuerit de nove[m]**
brachiis velocius volvitur / Et debet intus
dividi per colunnas tenentes eam in sex
partibus Et coppulate decte colunne poni
intus et in billicis suis debent esse buse
de ligno quercus aut illicis aut cedrangogli
quia minus consumantur et equaliter poni
cum quadrante. Et ista talis rota est acti-
bilis ad† omne ingienium molendinorum
et [h]auriendum aquam et ad volvendum
multas ruchas per torquendum sireum et
rephes. Et est utile edifitium quia ipsum
est unive[r]sale.

* The text has nove*um*, changed to nove[m]
by striking out u*m*.
† om*n*e*m*, preceding o*m*ne, has a line through
it.

The large wheel, shown in the design
on the other side, next to a deep well,
is good for drawing water with very
large pails which carry much water. If
its shaft is short it can be turned better
and more rapidly by the man treading it
inside. It should be fourteen feet high,
but if it measures eighteen feet, it can
be turned more rapidly. Inside it should
be divided by columns which hold it at
six points, and said columns should be
placed inside in pairs. In its bearings
there should be sleeves of oak, ilex or
cedar wood, because they suffer less
from wear and tear. They should be
placed evenly, with a builder's square.
The said wheel can be adapted to [the
driving of] any machine for mills and
water-raising, and to turning many
wheels for twisting silk and thread. It
is a useful device, as it is universal.

31. Siphon on a bridge **(III, 32r).**

*Vide pontem posteriorem iusta castrum
in isto folio / aqua surgiens iusta fossatum
cum ripis suis non potest transire sine
ponte et si fuerit costructus pons planus
o[b]duceretur ab impitu aque / oportet
ergo quod edificetur ibi pons ad arcum / ut
sub eo deluvium aque transire et velociter
fluire possit. Et postea instruantur canne
[in] superficiem pontis et bucca canne
su[r]gientis sit minor canna expluente per
lo[n]gitudinem / canna versans sit maior
in longitudine quarto pluri Et modum as-
sciendendi aquam super pontem require
exemplo secundo libro draconis a fo[lio]
96 et apertius apparet in primo libro leonis
a fo[lio] 69 manu decti Ser Mariani*

Behold the bridge in back, and near it
the city, as shown on this sheet. The
water, which springs forth next to the
river and its banks, cannot pass across
without a bridge. If the bridge were
constructed flat, it would be carried
away by the impact of water. Therefore
let a bridge be built in the shape of an
arch, so that a flood of water may pass
below it and flow off swiftly. Then let
pipes be installed on the bridge. Let
the intake of the rising pipe be smaller
in length than the discharge pipe. Make
the discharge pipe greater in length by
an extra one-fourth. To learn the way of
lifting water onto the bridge, turn to the
example in the Second Book, of the
Dragon, folio 96 [II 94v]; and it appears
more clearly in the First Book, of the
Lion, fol. 69 [I 73v], done by the hand of
said *Ser* Mariano. [See Plates 12, 18.]

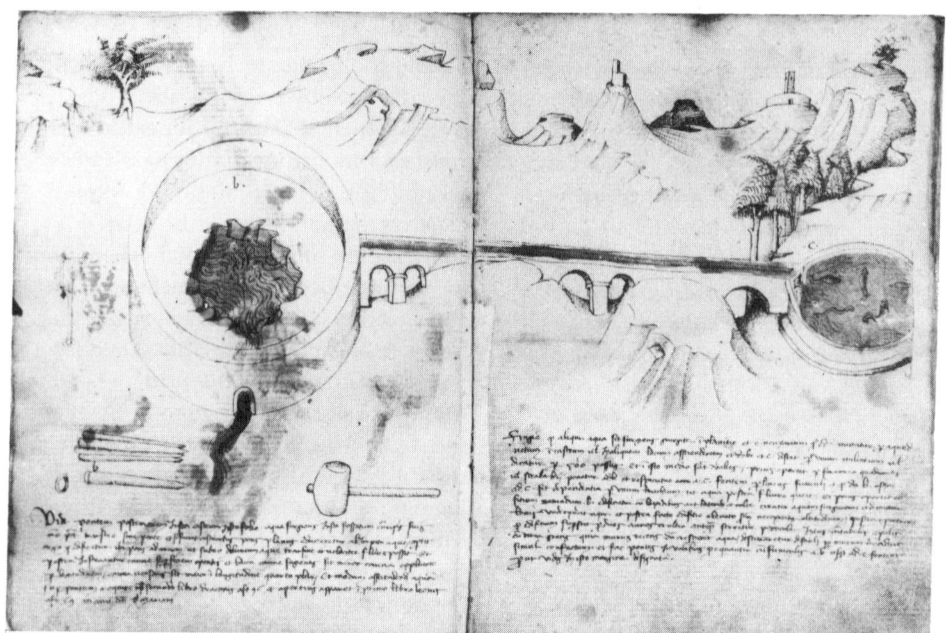

32. Text relating to the siphon **(III, 32v).**   33. Aqueduct and reservoirs **(III, 32v–33r).**

Fingie quod *aliqua aqua sit surgiens giur-*
*giter** *in planitie et* est *necexarium* quod
ipsa mictatur per aqueductum. in castrum
vel in aliquem locum assciendentem et
de .b. et c distat per unum miliarium
vel dicatur per 500 passus Et in isto medio
sunt valles / prius oportet per foramina
quadrantis vel stralabii ponatur ad b. et
respiciatur a .c. fontem† per *lineas funiculi*
et quod *da b. usque ad .c. sit dependentia*
per *unum brachium / ut aque postea fluire*
queit / et prius oportet edifitium retundum
.b. edificetur cum lapidibus aut lateribus
et calce circiter aquam surgientum et demic-
tetur locus unde exeat aqua et postea facto
edifitio elevato secundum necexitatis altitu-
dinem / postea oportuerit quod *edifitium*
superstet per duos annos et ultra / ante-
quam serretur porticula(m) intus cum
tabulis et palis cum terra gretis / quia
murus recens dum *cresceret aqua / destru-*
aretur de facili propter *murum non ad hic*
simul confactum. et fac pontes in valibus
que equantur cum funiculis a .b usque
ad .c. fontem prout vides in isto margine
designata.

* In the copybook Add. 34113 the words are
surgha surgha.
† Original text: *respiciatur a. c a / .c. fontem.*
The first two letters, *a. c,* are canceled.

Assume that some water filters from the
ground in a plain, that it is necessary
to bring it by aqueduct into a city or
other elevated place, and that from "b"
[a reservoir for the water] to "c" [a
fountain in the city] there is a distance
of one mile or five-hundred paces, with
valleys in between. First, through the
sightholes of a quadrant or astrolabe,
placed at "b," observe the line to "c,"
the fountain, along strings. Let there be
an incline of two feet from "b" to "c,"
so that in due course [a stream of] water
may flow [along it]. And first let a round
edifice "b" be built, of stone or brick and
lime, around the rising water, and let
there be a place where the water may
leave it. Then, when the edifice has been
built and raised to such height as necessary, thereafter let the edifice stand two
years or more, before the outlet is closed
from within with pile sheeting and clay.
For while the new wall rises it would be
easily destroyed by the water, as the
wall has not yet been able to bind. Build
bridges in the valleys, making them
straight by means of strings, from "b"
to "c," the fountain, as you see it designed on this sheet.

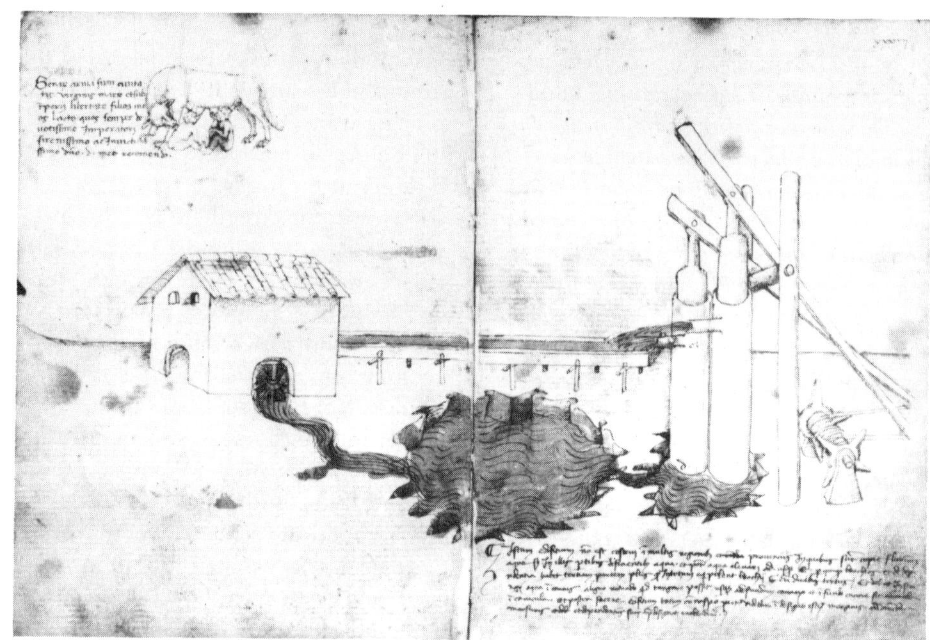

34. Waterpower system for a mill **(III, 33v–34r).**

Istum edifitium non est costrui in multis regionibus et provinciis in quibus sunt copie fluminum aque sed in illis partibus deficientibus aque. et potest aqua elevari a d. usque .e. per quinque brachios et de dependentia habet tertiam partem plus quod in totum equipolent brachii [6] cum duabus tertiis† / Et debet instringii aqua in cannis a ligno retundo quod tangare possit usque ad fundum cannarum et in fundo canne sit alimella in can[n]icula. Et postea facere edifitium totum cum naspo prout videtur in designo istius marginis cum omnibus mensuris § et dependentiis suis predesignati molendini etc*

* Some letters are canceled after *et*. Add. 34113 writes: *molti luoghi ne in molte provinccie.*
† Add. 34113: *braccia 5 2/3 [?].*
§ *et debet* is canceled before *et*.

This device is not to be constructed in those many regions and provinces where water-bearing rivers are copious, but in those parts that are poor in water. Water may be elevated [along an angle running] from "d" to "e" by ten feet. For incline, one-third is taken. Adding this, we have a total of [13] and one-third feet. Water must be forced in cylinders by a round piece of wood that can move to the bottom of the cylinders. In the bottom of each cylinder, let there be a flap valve in a small tube. Then complete the device with a winch, as seen in the design on this page, with its proper measurements and inclines for the above-designed mills.

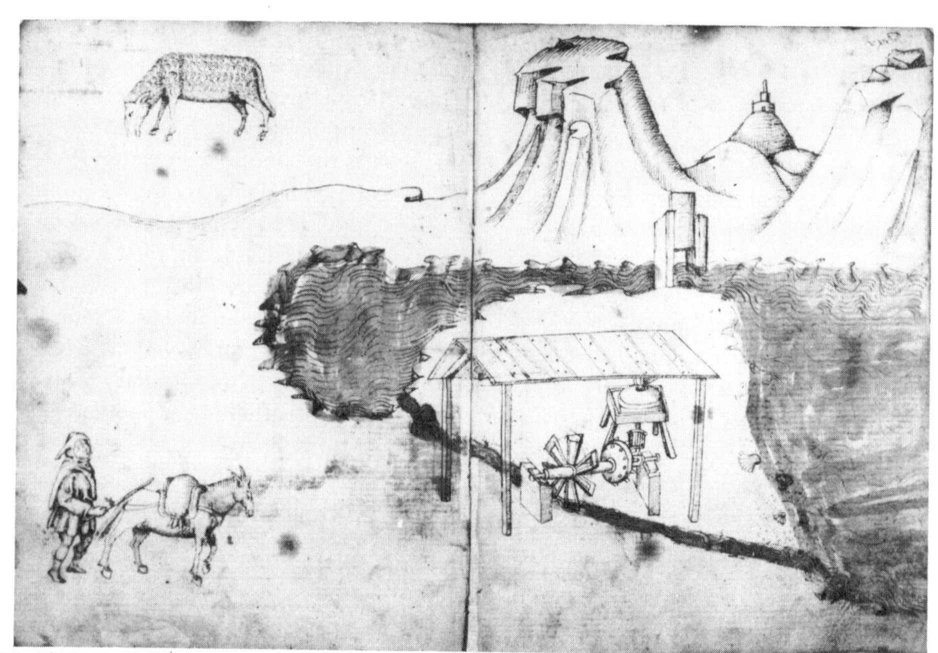

35. Tide mill **(III, 34v–35r).**

36. Text relating to the tide mill **(III, 35v).**

Fol. **35v** (Plate 36)

Edifitium super litera maris. costructum in posteriore parte istius pupirii oportet quod sit staneum naturale vel pelagus quis fluire et flui possit iusta mare vel secus flumina set melius est iusta mare et si staneum non est naturale artifitialiter fiat cum duobus itineribus ex quibus per unum intrat aqua in foveam mare crescente / et postea deficiente aqua maris serretur introitus cum cateracta et tunc oportet quod aperiatur alia pars exitus aque que vadit ad ruccas molendini prout cernitur in designo prenominato de molendino. et quando.†*

* In the manuscript appears *alias,* changed to *alia.*
† Add. 34113: *nello disegnio del mulino guidi sopra disegniato chon due chateratte.*

For the device on the seashore, constructed as shown on the other side of this sheet, let there be a natural pool or body of water which can have flux and reflux, near the ocean or along rivers but desirably near the ocean. If there is no natural pool let one be made artificially, with two connections, one of which allows water to enter into the pond when the sea rises. Thereafter, when water of the sea is low, let the inlet be closed by a sluice gate, and then let the other way be opened as an outlet. Water may then go to the wheels of a mill as seen in the above-named mill design. [These operations may be repeated] from time to time.

37. Reversible hoist **(III, 36r).**

38. Text relating to the reversible hoist **(III, 36v).**

*Ingienium magne rote que in posteriore
parte istius foleii apparet per ecum volvitur sciendum est quod ipsum [edificatur]
ex tribus causis primo quod ipsum velociter
giratur secundo / quod ipsum est facile ad
trhaendum propter quod magna pondera
altius levantur /. tertio quod ipsum prociedit et non retrociedit qu[a]rto non de-
mictitur tempus quod multum prevalet ad
edifitium magne fabrice Et quando rota
est magna cum dentibus suis lateralibus
facilius ille due parve ruche fisse in stilo
volvuntur cum festinatione. Et quando
animal firmatur propter pondus eventum
ad summitatem / tunc cum pertica volve
vitem super quam firmatur stilus magne
rote Et quando volvitur vitis sursum cista
alta venit deorsum et inphima vadit sup-
rossum. Et quando decta vitis volvitur
deorsum cista inferioris ass[c]iendit sursum / et tunc cista de sursum desciendit*
deorsum prout cospicitur in designo prenominato.*

* After *desciendit* appears *in*, which is canceled.

The Engine with the Great Wheel, which
appears on the other side of this sheet,
is turned by a horse. Let it be known
that it is built for three reasons. First,
because it turns rapidly. Second, because
it facilitates lifting, as large weights are
raised on high. Third, because it runs
forward and not backward. Fourth, it
does not waste time, which is most important in building a large structure. If
the wheel with lateral teeth is large, two
small pinions fixed on one shaft can
easily revolve with rapidity. When the
animal is stopped because the weight
has reached the top, then with a rod
turn the screw that supports the shaft
for [the pinions of] the Great Wheel.
When the screw has been turned up, the
upper basket comes down and the lower
one goes up; when the said screw has
been turned down, the lower basket
rises and then the basket from above
descends, as is seen in the aforesaid
design

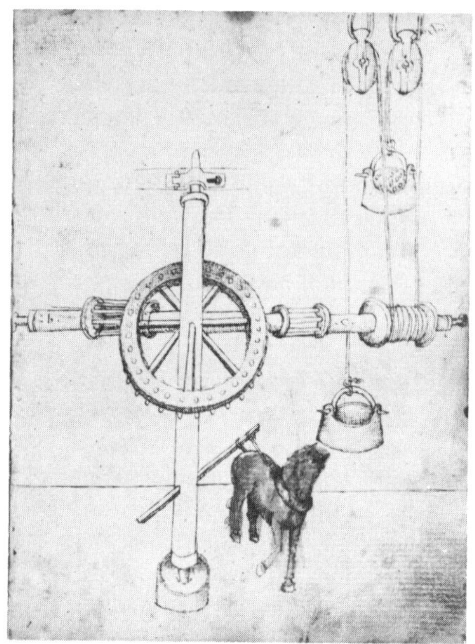

39. Reversible hoist (variant) **(III, 37r).**

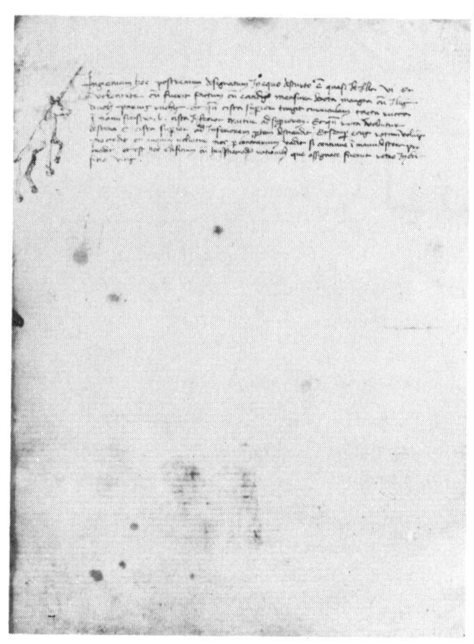

40. Text relating to the reversible hoist **(III, 37v).**

*Ingienium hoc postremum designatum
in equo descurto est quasi de illa vi Et
velocitate cum fuerit factum cum can-
dissimo mensura de rota mangna cum
illis duobus parvis ruchis. Et quando cista
superior tangit curriculam / tacta rucca
in manu sinistra .b. cista inferior tra[h]itur
ad superiorem. Et quando rucca volvitur
destera c. cista superior ad inferiorem
partem descendit / Et semper ecus rotam
volvens prociedit et nu[n]quam volvitur
nec per contrarium vadit sed continue in
manu destera prociedit / et est hoc edifitium cum huiusstie modi rationibus que
assignate fueru[n]t retro in edifitio vitis.*

The machine designed here on the other
side, which is run by a horse, has almost
the same power and velocity [as the
preceding one] if most accurate measurement is used for its Great Wheel, with
its pair of small pinions. When the upper
basket has reached the idler pulley, with
the pinion "b" on the left hand engaged,
the lower basket is drawn up. When the
right pinion "c" [then] is turned, the
upper basket descends to the lower
parts. The horse always goes forward
as he turns the wheel. He does not go
or turn in the other direction but con-
tinually proceeds to the right. The same
considerations apply to this device which
were assigned, in back, to the Device
with the Screw [36r, 36v].

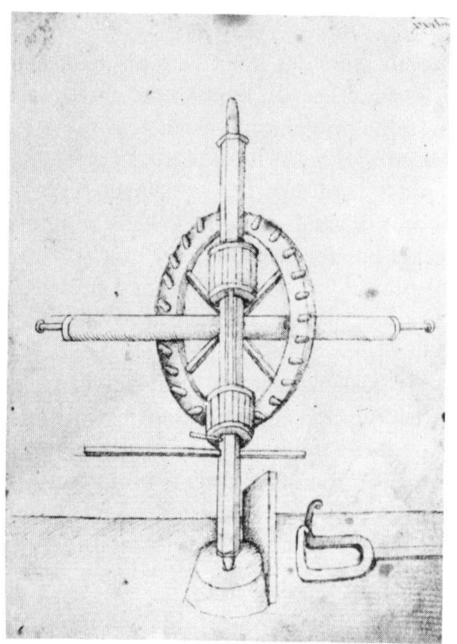

41. Reversible hoist (another variant)
(III, 38r).

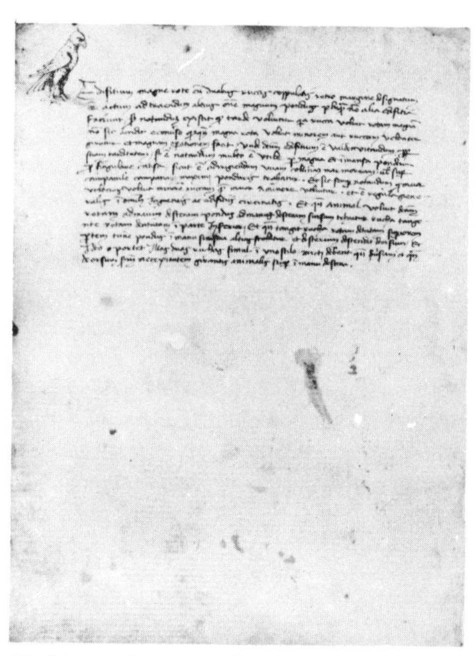

42. Text relating to the reversible hoist
(III, 38v).

*Edifitium magne rote cum duabus ruccis
coppulatis retro margine designatum est
actum ad tra[h]endum altius omne magnum pondus plu[s]quam non alia edi-
fitia faciunt / sed notandum existet quod
tarde volvitur quia rucca volvit rotam
magnam non sic accidit [si] e[st] comisso
quod (quando) magna rota volvit minorem
aut ruccam velociter giratur et magnam
operationem facit / unde dectum edifitium
est valde vitandum propter suam tardita-
tem / sed est notandum multo est utile
per magna Et immensa pondera per sin-
gulare[m] causa[m] / sicut est ad[d]iri-
giendum unam colunnum murmoream / vel
super campanile campanarum maximi
pondaris trahatur. Et sic semper notandum quod maior velocius volvit minorem
ruccam quam maior a minore volvitur.
et est regula gieneralis in omnibus in-
gieneis ac edifitiis circinatis. Et quando
animal volvit dectam rotam ad manum
desteram pondus ad manum desteram sursum trhaitur rucha tangente rotam dentatam in parte inferiori Et quando tangit
rucha rotam de[n]tatam superiorem partem tunc pondus in manu sinistra altius
scollitur / et desterum descendit deorsum /
Et ideo oportet illas duas ruchas simul in
uno stilo / micti debent quando sursum et
quando deorsum [sunt] secundum necexi-
tatem girantis animalis semper in manu
desteru [procedit].*

The Great Wheel Device with two coupled
pinions, as designed in the margin in
back, is suitable for drawing up any
great weight, greater than other devices
can handle, but let it be noted that it
turns slowly, since the [small] pinion
turns the Great Wheel. It is different
when the Great Wheel turns the smaller
wheel or pinion—it then rotates quickly
and does much work. Wherefore the device often is to be avoided because of
its slowness. But let it be noted that it
is very useful for great and immense
weights in particular cases, as in conducting a marble column of extreme
weight, or lifting bells onto a bell tower.
Be it always noted that the Greater Wheel
turns the smaller one more rapidly than
the Greater One is turned by the smaller.
This is a general rule in all machines
and devices with circular motion. As the
animal turns this wheel to the right, the
right-hand weight can be drawn up
while the pinion contacts the toothed
wheel in the lower part. When the pinion contacts the toothed wheel in the
upper part, the left-hand weight is lifted
up high and the right one descends
downward. Both pinions must be placed
on the one shaft. Whether they be up
or down, as may be required, the animal
turning them always proceeds to the
right.

43. Drive for a bucket chain **(III, 39r).**

Fol. **39v** (Plate 44)

Edifitium rote actibile actigiendum aquam
de cisterna prout retro indicatur per desig-
num in bove volvente / est in multis locis
longi[n]cuis Et propinquis / et quasi potest*
dici universale ingienium / et potest or-
dinari in duobus modis in primo. prout
patet per se in designio / in secundo ubi
dentes rote tangunt ruccam ex latere suctus
potest ordinari quod prefate rote dentes
tangant costas rucce ex latere superiori et
operantur illud idem et suaviter volvitur
rota et citius [h]auritur aqua cum vaxibus /
si rota est cum (in) billicis suctilibus Et
potest edificari rotam quinque brachiorum
per altitudinem et si minor fuerit minus
actigiendum operatur / et si maior erit
rota plus actingit propter plures dentes
continuos volgientes eam / Et notandum
existit quod [si] aliquando† vasa posita
cum funiculis super rotam non trahantur
ordinate / tunc oportet quod in aqua cis-
terne fiat alia rota minor bilicata in medio
finiculorum vasa tenentium.

* Add. 34113: *quali.*
† After *aliquando* appears *quod,* canceled.

The wheel device for obtaining water
from a cistern, as indicated in back by a
design with an ox turning it, is [used]
in many places, far and near. It can al
most be called a universal machine. It
can be fitted in two ways: first, as is
shown in the drawing; second, so that
the teeth of the wheel touch the pinion
downward, [that is], it can be fitted so
that the teeth of the aforesaid wheel
touch the ribs of the pinion from above.
They work this [pinion], and the wheel
turns smoothly. Water is drawn more
readily with containers if the wheel is
made with thin pivot pins. The wheel
can be made ten feet high. If it is smaller
it does less work. If it is larger it raises
more, because of a greater number of
consecutive teeth turning it. Let it be
noted that if containers, hanging on
ropes [which pass] over the wheel, are
not drawn properly at some time, then
it is best that another and smaller wheel
be made, which is pivoted between
ropes holding the containers, in the
water of the cistern.

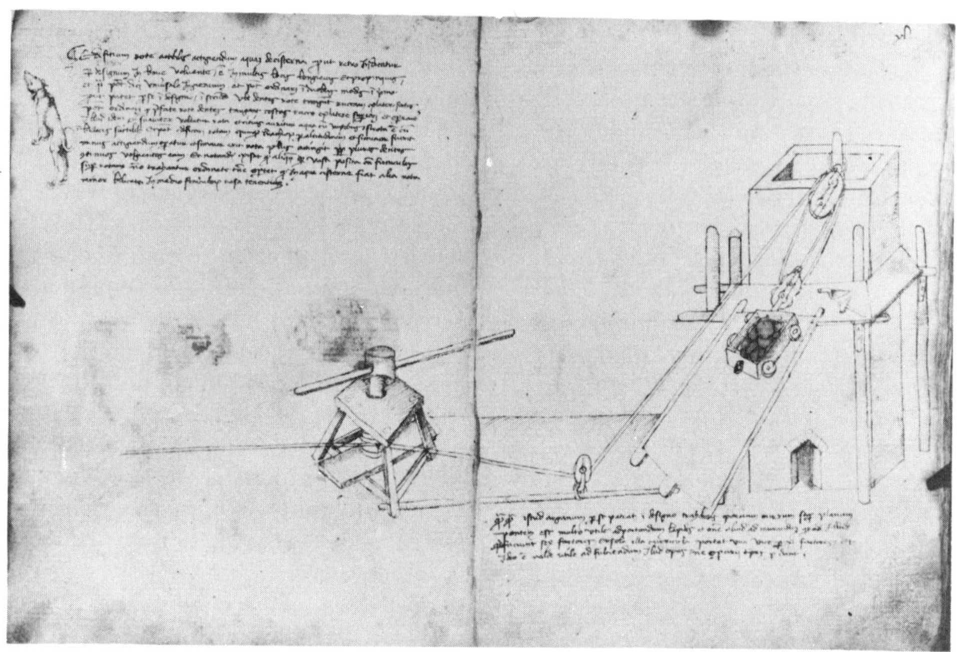

44. Text relating to the drive for a bucket chain **(III, 39v).**

45. Winch for a wagon **(III, 39v–40r).**

Fol. **40r** (Plate 45)

Quamquam istud arganum per se pateat in designo trahens parvum currum super planum pontem est multo utile ad portandum lapides et omne aliud ad murandum (quod) illud quod faciunt sex fartores / Et sola illa curricula portat una vice per X fartores / Et ideo est valde utile ad fulciendum il[l]ud opus / ne operarii tempus perdant.

Although it is obvious from the drawing how this winch draws a small cart up a flat bridge, it is very useful for the transport of stones and all other needs of masonry work. This simple buggy carries at one time that [quantity of material] that six laborers and [even] ten laborers use. Thus it is very useful for performing this work so that masters do not lose time.

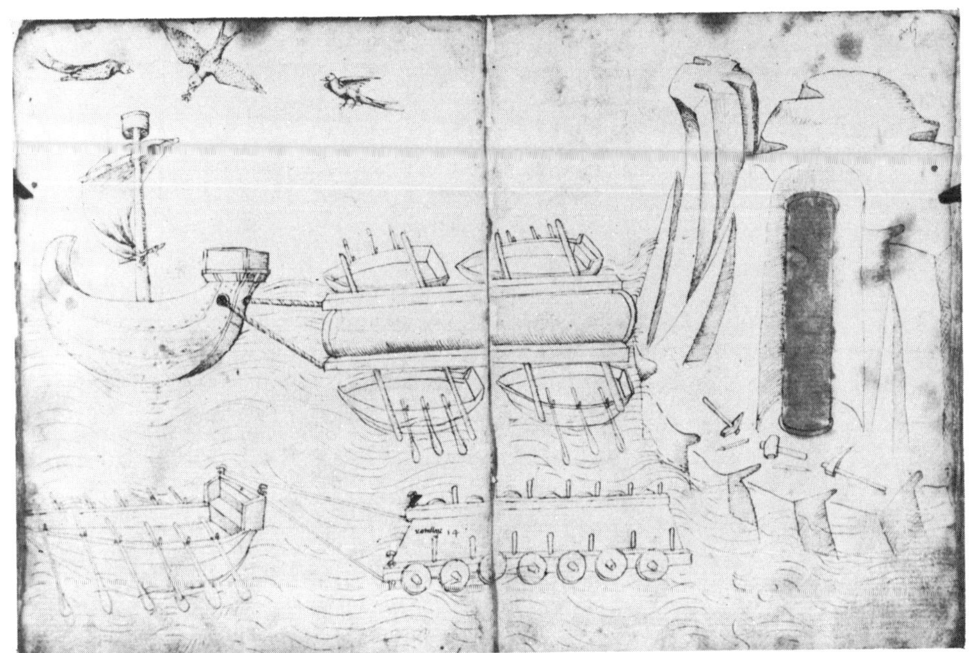

46. Quarrying and handling large columns **(III, 40v–41r).**

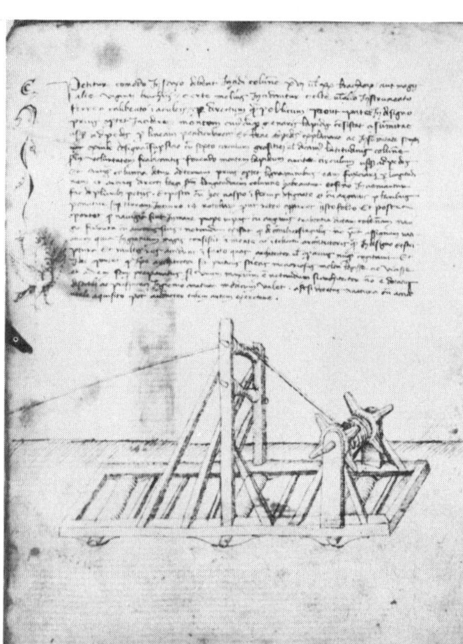

47. Text relating to quarrying. The architect's work. Winch and roller frame
(III, 41v).

Fol. **41v** (Plate 47)

*Petitur comodo in saxo debent incidi col-
unne XVI vel XX brachiorum / aut magis
alte viginti brachiis. / certe melius incidan-
tur cielte vel alio instrumento ferreo c[h]ali-
beato in aculeis /. per directum quam per
oblicum / prout patet in designo prius
oportet incidere montem cuiuscumque
generis lapidis consistat a summita[t]e
usque ad pedes per lineam [per]pendicu-
larem Et bene ad pedes explanare ac in
summitate saxi Et exinde designa in superficie cum sexto circulum grositiei vel dicamus latitudinis colunne secundum vol-
untatem frabicantis fovendo montem
lapideum circiter circulum usque ad pedes
Et antequam colunna detur ad terram
prius oportet lignaminibus eam faxciari
per longitudinem et arbores directi longi
secundum longitudinem colunne habeantur
Et si non inveniantur fac de pluribus petiis.
Et post[e]a cum hoc naspo inferius designato et cum arganis pluribus ponatur super
terram in curro 14 rotellarum prout retro
apparet in isto foleo Et postea oportet
quod navigia sint in mare prope ripas cum
arganis trahentia dictam colunnam navigia
firmata cum ancoris suis. notandum existit
quod de omnibus et singulis non potest
assignari rationem quia ingienium magis
consistit in mente ac intellectu architec-
toris quam in designo et scriptura Et multe
res accidunt in facto / quas architector
vel operarius nunquam cogitavit. Et ideo
oportet quod ipse architector sit pra[c]ticus
sciens menoriosus molta legisse ac vidisse
et ad rem semper preparatus / sed unum
maximum est notandum / si architector
non est dotatus de soctili ac prespicaci
ingienio a natura modicum valet. ast si
dotatus a natura cum accidentale aquisito
potest audacter talem artem exercere.*

The question is raised how columns of
30 feet or 40 feet, or more than forty
feet high, shall be cut out of rock. It is
surely better to cut them with a chisel or
other steel-edged iron tool into upright
needles [obelisks], as shown in the drawing, rather than in oblique position.
First let [test] cuts be made into a mountain to see of whatever kind of stone it
may consist, from top to bottom along
a perpendicular line. Let it be flattened
well at the bottom and top of the rock.
Then with a compass, design on its
surface a circle that gives the thickness
or, as we should say, the width of the
column, as the workman wishes. Hollow
out the stone mountain all around along
the circle, [down] to the bottom. Before
the column is lowered down to earth,
first let it be bound with timber along its
length, and let beams be directed along
the length of the column. If they cannot
be found [long enough], make them of
several pieces. Thereafter, with the wind-
lass shown below [41v] and with several
capstans, let it be laid on the ground,
on the wagon with 14 wheels, as appears
on the back of this sheet. Then have
vessels in the sea near the strand with
capstans to draw the said column. Make
the vessels firm by their anchors. And
let it be noted that one cannot explain
each and every detail, because ingenuity
resides in the mind and intelligence of
the architect rather than in drawing
and writing. Many things occur in the
course of the work that the architect or
worker never planned. Therefore let the
architect be experienced and learned. He
should have read a good deal and seen
much, and be always prepared. But one

principal thing is to be noted: if the
architect is not gifted by nature with
subtle and perspicacious ingenuity, he
is worth little, but if gifted by nature
with the forces of power over chance,
he can practice this art with confidence.

48. Cofferdam method of laying foundations (III, 42r).

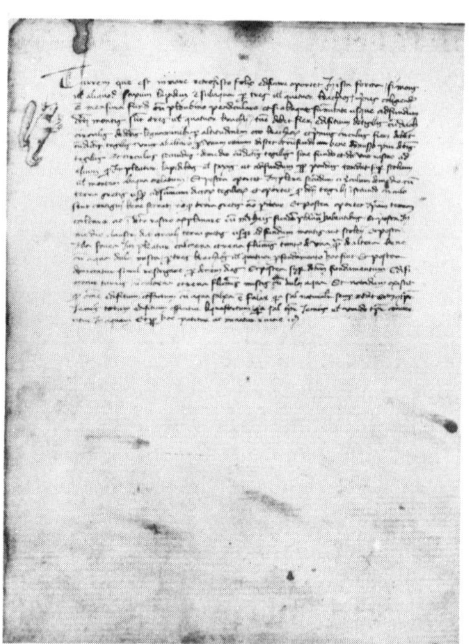

49. Text relating to the cofferdam method (III, 42v).

Fol. **42v** (Plate 49)

Turrem que est in mare retro in isto folio edificare oportet in ista forma / si mons vel aliquod saxum lapideum est sub aqua per tres vel quatuor brachios / prius coligiendum est mensura fundi cum plonbino [per]pendiculare / et si ab aque summitate usque ad fundum decti montis sunt tres vel quatuor brachii / tunc debet fieri edifitium de tigillis cum duobus circulis de dectis legnaminibus / per altitudinem otto brachiorum et primus circulus fieri debet cum dectis tegillis unus ab altaro per unum otta[v]um [brachii] distet et cum fundo bene commisso purdectis† tegillis Et circulus secundus de medio cum dictis tegillis*

* After *fundo* appear some letters, which are canceled.
† Add. 34113: *pure di detti.* Palat. 767: *pur con decti.*

A tower in the sea, as shown on the back of this sheet, is to be built in the following way. If there is a mountain or some stony rock six or eight feet below water, then first take measurements of the bottom with a perpendicular plumb line. If the distance from the water level to the floor on this mountain is six or eight feet, then a structure of sheeting timbers is to be built, in two circles thereof, to a height of sixteen feet. The first circle is to be built with one of said sheeting timbers distant from the other by three inches. A bottom is well connected to said sheeting. The second circle inside [the first] is built with similar sheeting but without a bottom. Between one gridwork and the other let stones or rocks be

[*sit*] *sine fundo et da uno rastro ad alium*
quod *impleatur lapidibus vel saxis / ut*
ad fundum propter *pondus tendat super*
scolium vel montem ab aqua celatum / Et
postea oportet implere secundum circulum
de medio cum terra gretis usque ad sum-
mum dictorum tegiliorum et oportet quod
decti tegilii in secundo circulo sint contigui
bene serrati / eo quod terra gretis non
perdetur / Et postea oportet ipsam terram
calcare ac in decto rastro explanare cum
*matheis fund*um *planum habentibus / Et*
post[e]a in medio claustri dicti circuli [erit]
terra gretis usque ad fundum montis aut
scoleii / Et post[e]a illa fovea impleatur
calcenu et rena fluminis tantum de una*
quam *de altera bene cum aqua dulci mista /*
per tres brachios vel quatuor per fonda-
mentu huc fiat Et postea demictatur simul
restri[n]gare per decem dies†/ Et postea
super dectum fundamentum / Edificetur
turris cum calcena et rena fluminis mistis
cum dulci aqua / Et nota[n]dum existit
quod *omne edifitium confectum cum aqua*
salxa est fal[l]ax quia *sal naturaliter sem-*
per rodit Et in tempore *[h]iemis totum*
edifitium efficitur liquefactum / quia sal
tempore [h]iemis vel [h]umido tempore
convertitur in aquam Et propter *hoc patitur*
ac minatur ruine i[n]tus.

* *calciena* changed to *calcena*. Add. 34113:
calcina.
† Add. 34113: *per .10 o .12. di.*

filled in, that their weight must cause
the structure to sink to the bottom on
the rock or mountain hidden below
water. Thereafter let the second or inner
circle be filled with clay to the top of
said sheeting. Let the said sheeting of
the second circle be contiguous, well
sealed, so that the clay will not seep out.
Thereafter let said clay be rammed and
flattened inside said gridwork with
flat-ended mallets. Then on the inside
of said circle there will be [rammed]
clay, down to the floor on the mountain
or cliff. Then fill this pit with lime and
river sand, as much of the one as of the
other, well mixed with fresh water, six
or eight feet deep. In this way the foundation is to be built. Then leave it all to
bind for ten days. Thereafter the tower
can be built on this foundation, with
lime and river sand mixed with fresh
water. And be it noted that any building
constructed with salt water is faulty,
because salt by nature always disinte-
grates. In wintertime the entire structure
would dissolve, because salt converts to
water in time of winter or rain, wherefore [the structure] suffers and is threat-
ened by destruction from within.

50. Box caisson process **(III, 43r)**.

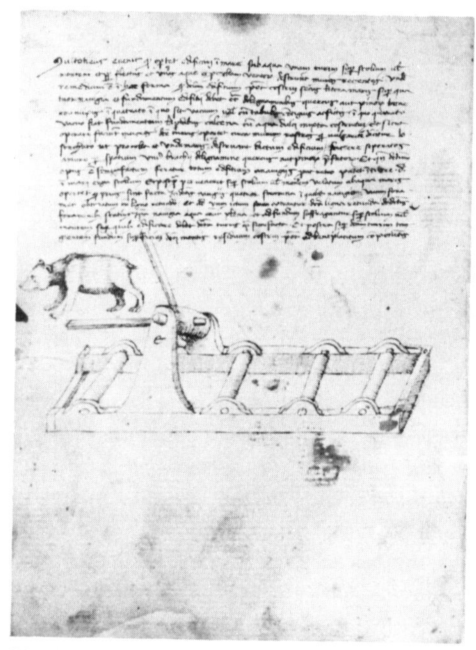

51. Text relating to the box caisson process **(III, 43v)**.

Multotiens evenit quod oportet edificari in mare sub aqua unam tur[r]im super scolium vel montem et propter flettus et undas aque et procellam ventorum / destru[u]nt muros recentes / unde remedium est in hac forma / quod dictum edifitium potest costrui secus litora maris super quatuor navigia et fundamentum edifitii debet esse de lignaminibus quercus aut pinorum bene conmixis in quatrato in quo sit vacuum val cum tabulis contiguis ac fictis / in quo quatrato vacuo fiat fundamentum lapidibus calcena cum aqua dulci mixtu costructum / Et sicut operarii faciunt mu(o)ros dicte turris / oportet circa murum rastros quae vulgariter dicitur lo slechato ut procella et unda maris ne destruant dictum edifitium / faciere seperatos a muro per spatium unius brachii de lignamine quercus aut pinorum prefatorum. Et quando dectum opus est semifactum / feratur totum edifitium a navigiis prout retro patet in turre .d. in mari erga scolium Et postquam pervenerit super scolium vel montem velatum ab aqua maris oportet quod prius sint facta in dectis navigiis quatuor foramina in quolibet navigio† unum foramen obturatum cum ligno retundo / Et ad un[u]m ictum§ extrahantur. decta ligna retunda de dectis foraminibus / statim ipsa navigia aqua erunt plena / et ad fundum suffragantur‡ super scolium vel montem super quibus edificare debet decta turris quasi semifacta. Et postea super dectam turrim tangentem fundum superficiem decti montis residuum costrui potest ad bene placitum componentis*

**vel* is obviously meant. Add. 34113: *overamente.*

† After *navigio* a letter is erased.

§ After *ictum* a word or syllable, perhaps *foram,* is canceled.

It happens often that a tower must be built at sea, on a cliff or mountain below water, and that spray, waves or windstorms destroy the recently constructed walls. For which the following is a remedy. Said structure can be constructed near the strand of the sea on four ships. Its foundation should be of oak or pine wood, well joined, and forming a square, upon which there should be provided a hollow space by contiguous and fitted boards. In this square, hollow space, let a foundation be built of rubble mixed with lime and fresh water. As the workers build the walls of said tower, let them put piling, commonly called palisades, around the walls, so that storm and sea waves may not destroy said structure. Make them of oak or pine wood as aforesaid, and distant from the wall by the space of two feet. When said work is half-built, let the entire structure be carried by the ships— as shown in back for the tower marked "d"—over the sea, out to the cliff. When it has arrived above the cliff or mountain concealed by seawater—four holes having been made beforehand, in the said ships, one hole in each ship, closed by a wooden plug— [then,] at a single stroke, let said wooden plugs be removed from said holes. Promptly said ships will be full of water and will sink to the floor on said cliff or mountain, where the more or less half-built tower is to be built [to completion]. Thereafter, when the tower rests on the floor, the surface of said mountain, the remainder can be constructed at the builder's convenience.

‡ *suffundantur* is probably meant.

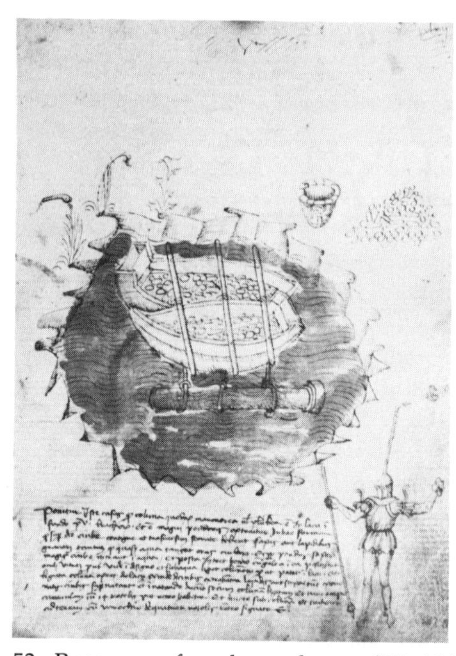

52. Recovery of sunken columns **(III, 44r).**

Fol. **44r** (Plate 52)

Ponitur iste casus quod colu[n]na quedam marmorea vel plumbea est in lacu in fundo XV. brachiorum. Et est magni pondoris / extraitur in hac forma quod superdecté cinbe contigue et trasversim serrate debent saxis aut lapidibus gravari tantum quod quasi aqua tangat oras cinbarum. Et propter pondus saxorum magis cinbe intrant in aqua[m] / Et post[e]a intret homo cum galea in ea positi sint oculi vitrei prout vide[s] in designo et sub aqua liget colunnam prout patet in lacu. Et ligata colunna exeat de lacu / exinde de cinbis extrahantur lapides aut saxa / tunc exoneratis cinbis supernatant et in natando ducunt secum colunam ligatam / Et tunc accipe curriculum cum 14 rotelis prout retro habetur. Et nucte sub colunna Et trahetur ad terram cum varochio de quatuor rotelis retro signato e.

The case arises that a column of marble or [pipe of] lead lies on the bottom of a lake, some thirty feet down. Such an object of great weight is raised in this way: the aforesaid vessels [III 43], arranged contiguously and connected sidewise, must be loaded with rocks or stones, to such extent that water almost reaches the top edges of the vessels. The more weight there is of rocks, the more deeply the vessels enter into water. Then let a man wearing a helmet with glass-eye [windows], as shown in the drawing, descend into the water and let him tie the column [to the vessels], wherever he may find it in the lake. When the column has been attached he may leave the lake. Thereafter let the stones or rocks be removed from the vessels. Then, as the vessels are unloaded, they are buoyed up, and as they rise they take with them the attached column. Then take the wagon with 14 wheels as shown in back [40v–41ı] to a position right under the column, and draw it to land by means of the winch with four rollers, [shown] in back [43v], marked "e."

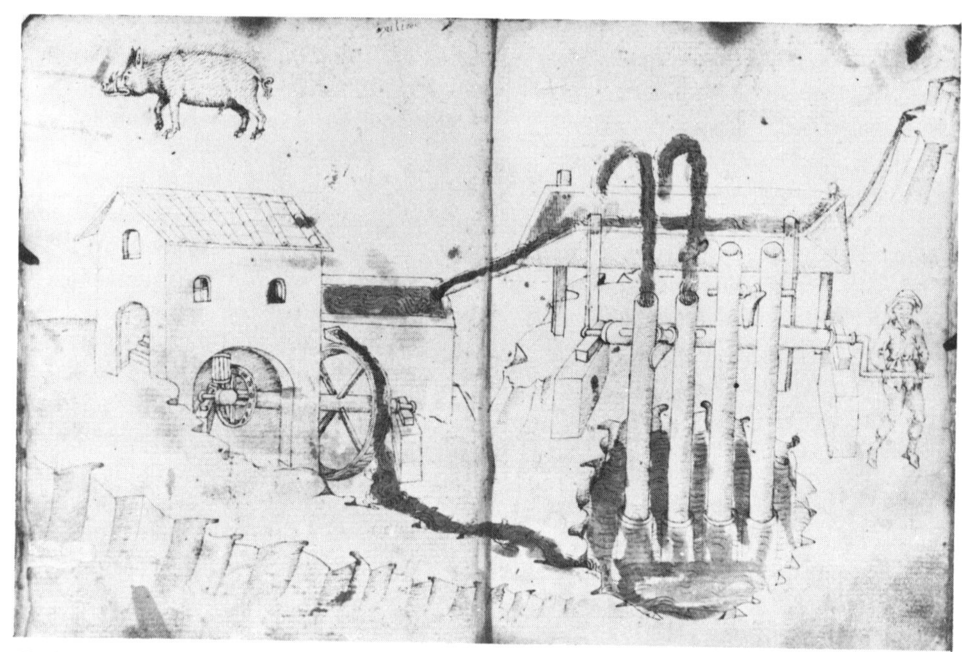

53. Water supply system for mills **(III, 44v–45r).**

54. Text relating to the water supply system **(III, 45v).**

Edifitium molarum quod retro videtur /
potest facere arch[i]tettor in duobus modis
ad rucham prope terram quod vulgariter
dicitur mulini terraguoli Et molti dicono
mulini ad reticino / et melius ac velocius
mola triticat cum .XX. palis quam XVI
Et plus satis cum XXIIII° quam XX / quia
aqua fluens et percutens in retecino in
summitate palarum propter maiorem re-
tunditatem velocius volvitur ac etiam non
dimictit tempus / propter palas contiguas /
aqua non interpellat /. Et in secundo modo
est dictum faciendum edifitium ad rotas
magnas in superficie cum capixectis / re-
cipientibus aquam de aqueducto cadentem
*Et istum molendinum vocatur galligenum**
eo quod est secundum mores francigen-
arum / et quando habet rotam de nove[m]
brachiis melius volvit ac velocius in se /
sed notandum est quod molendinum ad
retecinam velocius giratur quam molen-
dinum francigenum vel† gallicum dicatur.
Et oportet quod ille canne sugentes aquam
sint§ agentes et patientes de metallo quod
dicitur stagno cum animellis ab utraque
parte oppositis una alteri / Et debent dicte
canne agentes et patientes bene simul
com[m]isse per modum quod de sursum et
deorsum trasferantur cum rota ordinata
s[us]tollente dectas cannas prout cernitur‡
retro in designo de homine volgente stilum
dectarum cannarum etc Et istum edifitium
est valde bonum in illis locis ex quibus
deficiunt aque quia si est pelugus aqua
exire et redire in eum potest / in decto
designo apparet satis manifeste.

* In the manuscript appears *galligenus,* changed
to *galligenum.*
† After *vel,* three letters are erased.
§ Before *agentes* appears *agie. . . ,* which is
canceled
‡ In the manuscript appears *ciernitur,* changed
to *cernitur.*

The mill structure seen in back can be
built by the architect in two ways.
[Firstly,] with a small wheel near the
ground, then it is commonly called a
country mill, and many call it an open
mill. The millstone grinds faster and
better when there are 20 blades [on the
wheel] than when there are 16, and still
better with 24 than 20, because the water
that flows and hits the top of the blades
turns it faster because of its greater
roundness. Also, time will then be saved.
Because of contiguity of blades the water
will not strike between them. According
to the second method such a structure
is to be built with wheels of large cir-
cumferential area, having cups to receive the water that falls from the aqueduct. This mill is called Gallican, as it
is made according to French custom.
When the wheel has [a diameter of]
eighteen feet, it turns rather well and
rapidly by itself, but let it be noted that
the mill with open wheel turns more
rapidly than the French or Gallican mill.
Let the tubes for suction of water—
both those in motion and those at rest—
be made of the metal called tin, and with
flap valves on opposite ends. These
moving and resting tubes must be well
fitted together, in such a way that they
can be lifted and dropped by the wheel
provided for such lifting, as will be
seen on the drawing in back, where a
man turns the shaft for the said moving
tubes etc. This device is very good in
those places where water is scarce, because, when a pool has been formed,
water can be drawn off and returned to
it, as will be obvious enough from said
drawing.

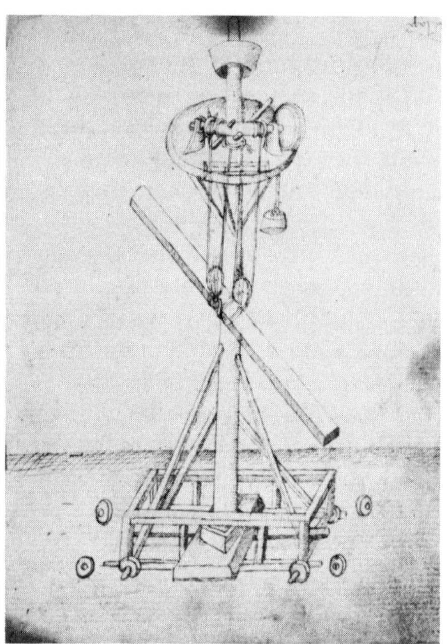

55. Builder's crane **(III, 46r).**

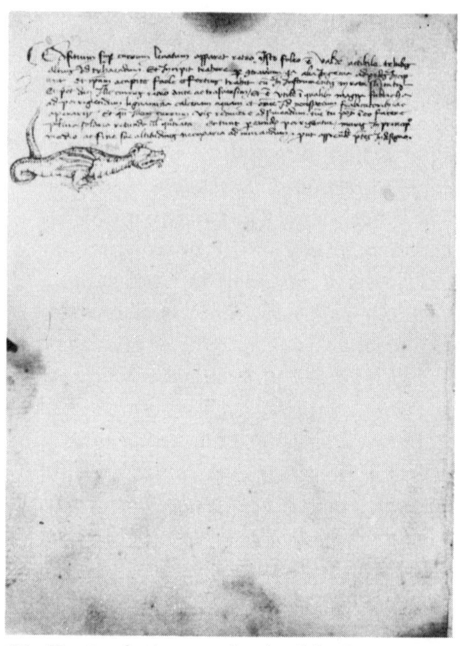

56. Text relating to the builder's crane. **(III, 46v).**

*Edifitium super currum levatum apparet
retro in isto foleo est valde actibile trhabes
altius ad trhaendum. Et incipit trahere
per contrarium / quia alia ingentu ud pedes
incip[i]unt Et ipsum a capite facile et for-
tius trahit / cum (in) instrumentis in rota
sum[m]itatis Et potest duci ille currus
retro ante ac trasversim / Et est utile in
qualibet magna fabricha ad porrigiendum
lignamina calcenam aquam et omnie id
necessarium frabicatoribus ac operariis /
Et quando illum currum vis reducere ad
serviendum. tunc tu potes in eo facere
plura solaria retundam vel quatrata / Et
tunc per eundem porriguntur / muris in
princip[i]o medio / ac fine sue altitudinis
necexaria ad murandum /. prout appren-
dere potes in designo.*

The device erected on a wagon, which
appears on the back of this sheet, is
very effective for the lifting of beams.
It applies the lifting effort in a way opposite to that of other machines, which
do it at the bottom. It does the lifting
easily and rather powerfully from the
top, by the wheel instruments in the
upper part. The wagon can be moved
backward, forward, and sidewise. In
any large construction work it is useful
for placing beams, lime, water, and all
that which is needed by the masters and
workmen. If you wish to make this wagon
serviceable, you can make several plat-
forms on it, either round or square. The
materials necessary for the wall masonry can then be placed on the walls
at lower, middle, and upper height, as
you may understand from the drawing.

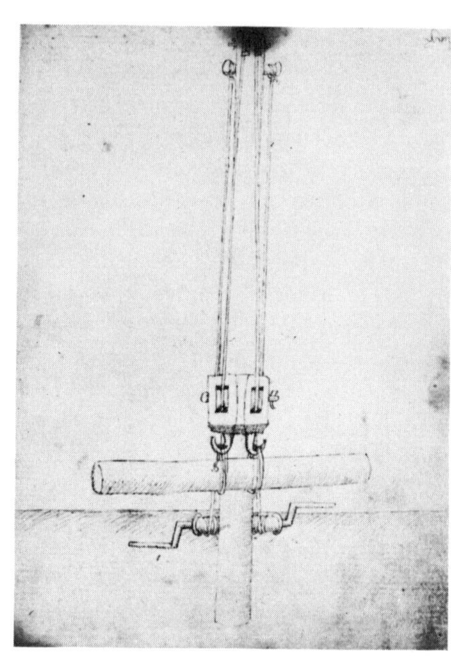

57. Another builder's crane **(III, 47r)**.

58. Text relating to the builder's crane **(III, 47v).**

59. Mill in a mountain landscape **(III, 47v–48r).**

Fol. **47v** (Plate 58)

Ingenium super arborem tonsum ac altum per directum est multo actibile ad levandum pondus magnum de terra et altius ducatur Et per se patet retro in isto folio / Item potest pondus super eum trahi cum duabus rotis in vice illorum manubreorum. Et dicte rote fieri debent prout patet in libro leonis a fo[lio] 16 ex manu meii Ser Mariani Iacobi dicti Tacchola. de Senis.

The machine on a high vertical post of cut lumber is very effective for lifting large weights from the ground and elevating them to the top. It is self-evident [from the design] on the back of this sheet. A load can be raised on it by two wheels, instead of the handles shown. Such wheels should be made as is shown in the Book of the Lion, fol. 16 [I 15], done by me, *Ser* Mariano di Jacopo, called Taccola, of Siena. [See Plate 2]

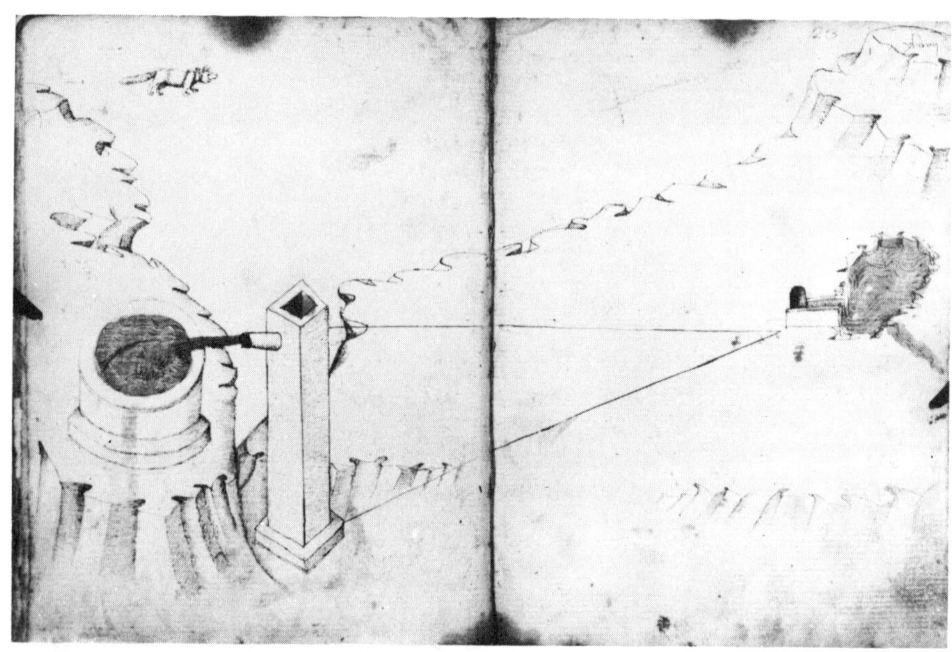

60. Water main and outlet **(III, 48v–49r).**

61. Text relating to the water main and outlet **(III, 49v).**

Fol. **49v** (Plate 61)

*Edifitium alte turris aquam expluentis per
se ipsum est satis manifestum quod aqua
naturaliter tantum assciendit quantum
desciendit Et ideo si mictatur per aqueductum aut per tunbam / oportet quod
ipsa tunba sit valde costructa bona calcena
cum rena fluminis et murus eius grossus
per unum brachium saltem / dicta tunba
murata intus in qualibet parte sui equaliter /
per altitudinem mensu(a)ra unius hominis
stature et per latitudinem duorum brachiorum / Et hoc sit pro duobus causiis /
prima quando mictitur aqua possit exalare
melius et oportet quod aqua turetur*
moderatim donec impleatur tunba / Et 2ª
causa est quando tunba est replena luto
et fimu quod agilius ipsa possit vacuari.
Et costrui debent prefata cum lineis et
dependentiis suis prout patet in designo
prefate. alte turris et infra c.*

* The manuscript shows *te'tur*. A translation,
in Add. 34113, is based on the reading *datur*.
However, we believe Mariano intended to
write *t'etur*.

High-tower structure for discharging
water. It is fairly obvious by itself that
water naturally rises to such height as
it has fallen. This applies if it is con
ducted through an aqueduct or a buried
conduit. Let the buried conduit be well
constructed of good lime with river
sand, and let its wall be at least two feet
thick. In each part of said buried conduit
the inside wall shall be equal in height
to the measure of a man's stature, and
be four feet in width. This should be
done for two reasons: firstly that the
water, when put [in the conduit], be
able to exhale better; the water should
be admitted gradually until the buried
conduit is filled. The second reason is
that when the buried conduit is full of
mud and dirt, it can be cleared out more
easily. It should be constructed with its
lines and incline as appears in the design of the aforesaid high tower, and
below at "c."

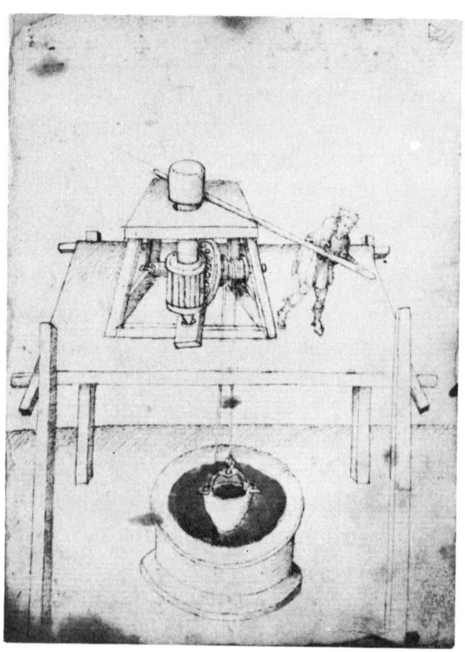

62. Crane **(III, 50r).**

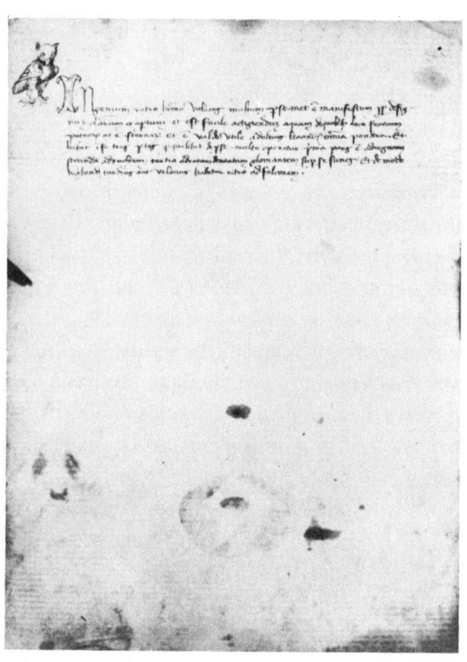

63. Text relating to the crane **(III, 50v).**

Fol. **50v** (Plate 63)

*Ingenium retro huomo volvens multum
per semet est manifestum propter desig-
num clarum et apertum et est facile actigiendum aquam de quolibet loco fontium
puteorum ac cisternarum Et est valde
utile ad altius levandum omnia pondera.
Et habet in se tres partes quod quo[d]libet
de per se multo operatur prima pars est
ad arganum secunda ad rucham / tertia
ad rotam dentatam glom[er]antem super
se funes / Et de modo trahendi tardius aut
velocius habetur retro ad falconem.*

The machine in back, with the man
turning it, is very obvious by itself from
the clear and straightforward design. It
is easy to use for obtaining water any-
where from fountains, wells, or cisterns.
It is very useful for lifting any weight.
It comprises three parts, each of which
is in much use by itself. The first part
is the capstan, the second the pinion,
and the third the cogwheel which winds
ropes onto itself. About the means of
drawing [up] more slowly or more rapidly, refer back to [the page of] the
falcon [III 38].

64. Crane (variant) **(III, 51r).**

65. Text relating to the crane **(III, 51v).**

Fol. **51v** (Plate 65)

Arganum retro trahens super lignum mag-
num canest[r]um cum duabus carriculis
habentibus duas rotellas pro qualibet dc
per se cum homine volgiente est valde
bonum ad omne pondus altius s[us]tollen-
dum. Et presertim bonum existit quando
trahitur lignum magnum ad tectum facien-
dum / tam in laribus / ec[c]lesiis ac tem-
plis.

The capstan in back, which draws a
basket up a high wooden post by means
of two blocks, each containing two pulleys, and by the turning effort of a man,
is very good for lifting any weight. It is
particularly good when a large timber
is to be raised to the roof of a large
shrine, church, or temple.

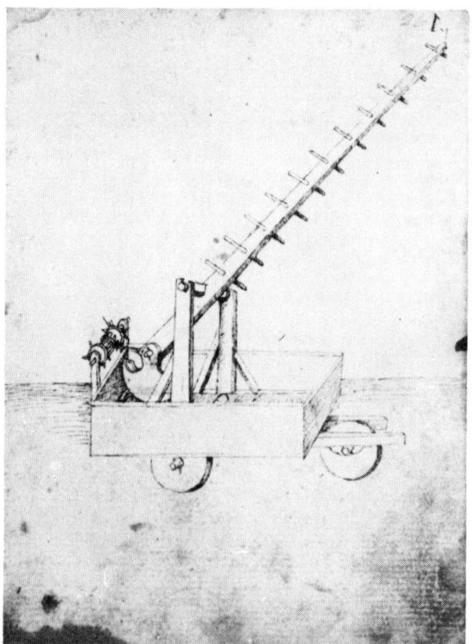

66. Ladder **(III, 52r)**.

67. Text relating to the ladder. **(III, 52v)**.

Fol. **52v** (Plate 67)

Vide currum retro habentem ascendiculum in medio ad i[n]star unius machine signate mangani. Est valde utile in edificatione unius templi / ad serviendum operaris ac magistris / sicut est aligandum et solvendum funes / pro mercenario stare potest in edifitio toc[t]ius fabrice quia ipsum actibile multum existit sursum et deorsum mictendum ad bene placitum frabicantium / prout latius in designo apparet.

In back you see a wagon which has a ladder thereon, similar to a certain machine inscribed "mangonel" [IV 66v]. It is very useful in building a temple, where it serves workers and masters, as it also serves soldiers in attaching and detaching ropes. It can be used in building any structure, as it is very workable for lifting and lowering and can be placed at the convenience of builders, as more fully appears from the design.

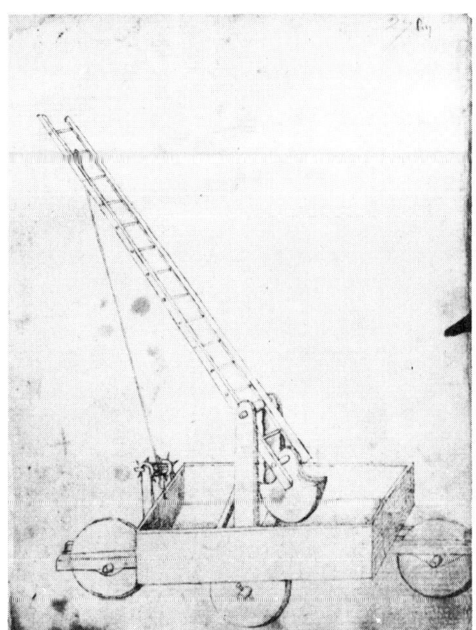

68. Ladder (variant) **(III, 53r).**

69. Text relating to the ladder **(III, 53v).**

70. Amphibious vehicle **(III, 53v–54r).**

Retro currus habens scalas el[e]vatas in se ac super se est conformus retro cum scimia et multo est utile in templo edifi-cando / in simili modo et formu cum scimiu etc

In back is a wagon with elevated scales thereon. Its upper portion is the same as that in back with the monkey [III 52]. It is very useful for building a temple, as is that with said monkey, etc.

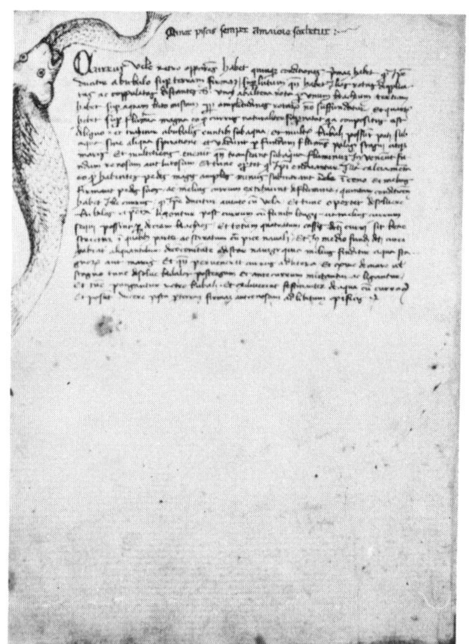

71. Text relating to the amphibious vehicle **(III, 54v).**

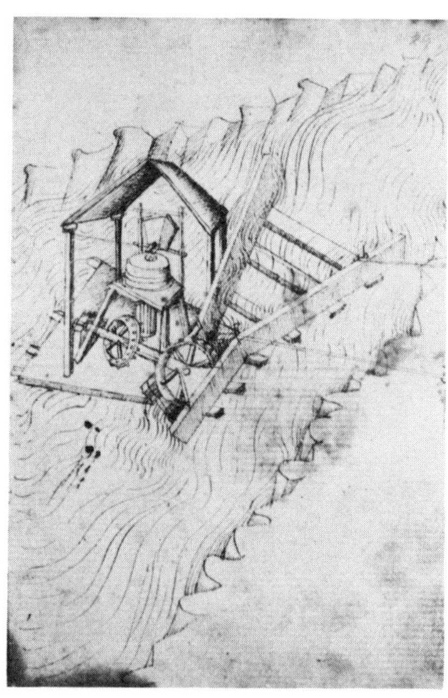

72. Floating mill **(III, 56r).**

Fol. **54v** (Plate 71)

Minor piscis semper a maiore sorbetur:—

*Currus vele retro apparens habet quinque
conditiones / primas habet quod ipse ducitur a bubalo super terram firmam / [al-
teram] super lutum quoniam habet illas
rotas dupplicatas ac coppulatas distantes
separatus unam ab altera rota per unum
brachium / tertiam habet super aquam luto
mistam propter amplitudines rotarum non
suffundatur / Et quartam habet super flumina magna eo quod currus naturaliter
supernatat quia conpositus est de ligno
et trahitur a bubalis euntibus sub aqua /
et multi bubali possunt pati sub aqua sine
aliqua spiratione et vadunt per fundum
fluminis palagii stagni atque maris / Et
multotiens evenit quando transeunt sub
aqua fluminum inveneunt fundum renosum*

The small fish is always swallowed by
the larger one.

The sail wagon which appears in back
has five conditions [under which it acts].
It has a first: that it is drawn by an ox on
firm ground. [A second:] on mud, for
which purpose it has double-spaced but
interconnected wheels, one wheel sep-
arated from the other by two feet. It has
a third: on water mixed with mud. Due
to the width of the wheels it will not
sink. Then it has a fourth: on large
rivers. The wagon naturally floats as it
is built of wood. It can be drawn by oxen
who walk under water. Many oxen can
stand being under water without breath-
ing. They walk on the bottom of the
river, lake, swamp, or sea. Often it hap-

*aut lutosum / Et tunc oportet quod ipsi
enduantur illa calciamenta eo quod habentes pedes magis amplos minus subintrant* in ceno / Et melius firmant pedes
suos / ac melius currum extrhaunt de
flumine / quintam conditionem habet ille
currus / quod ipse ducitur a vento cum
vela / Et tunc oportet desolvere bubalos /
et postea ligentur post currum cum funibus longis ut melius currum sequi possint /
per deciem† brachios / Et totum quatra-
tum cassis dicti curri sit bene strictum in
qualibet parte ac serratum cum pice navali /
Et in medio fundi decti curri habeat aliquantulum de concavitate ad i[n]star navigii
quia melius finditur aqua stagnorum aut
maris / Et quando pervenerit currus ad
litora / Et exire de mare vel stagno tunc
desolve bubalos postergum Et ante currum
mictantur ac ligantur / Et tunc pungantur
retro bubali. Et ex[s]iliverint festinanter
de aqua cum curro dicto Et pos[s]unt ducere
post[e]a per terram firmam aut cenosam
ad libitum opificis eius.*

* After *subintrant* appears *inde,* which is canceled.
† The *i* in *deciem* seems partly erased.

pens that, walking in river water, they
find sandy or muddy ground. Then this
footwear must be put on them, because,
when they have wider feet, they enter
less into the mud, set their feet more
firmly, and draw the wagon better from
the river. The wagon has, as a fifth
working condition, that the wind pro-
pels it by means of a sail. At such times,
let the oxen be detached from the front
and attached to the back of the wagon
with long ropes, so that they may follow
the wagon better at a twenty-foot distance. The entire square chassis of the
wagon should be tight in all parts, and
sealed with naval pitch. In its middle
let the bottom of the wagon have a slight
concavity like a ship, whereby it better
cleaves the water of the swamp or sea.
And when the wagon comes to shore and
[you wish] to leave the sea or swamp,
detach the oxen from the back and let
them be put and tied in front of the
wagon. Then let the oxen be prodded
in back, and they will quickly jump
from the water with said wagon. They
can follow their way on firm or muddy
ground, as the master wishes.

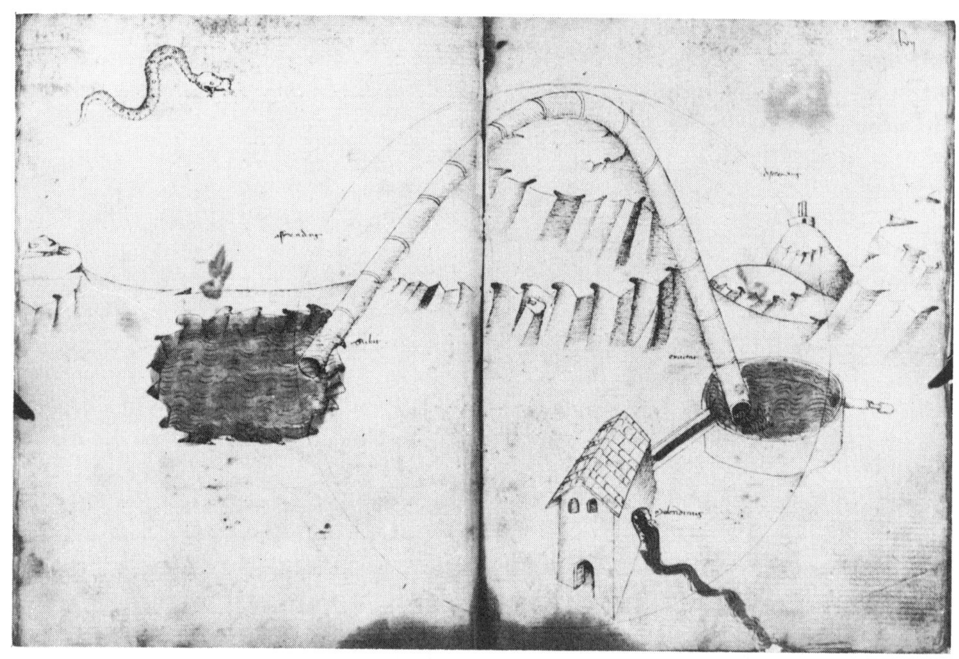

73. Siphon as water supply for a mill
(III, 56v–57r).

74. Text relating to the siphon **(III, 57v).**

Fol. **57v** (Plate 74)

Ingienium apparens retro in canna asscen-
dente super montem est valde amirabile /
quod turato in aqua primo motu semper
aqua asscendit et descendit / Et oportet
(quod) primo quod descendentia sit maior
quarta parte asscendentia / Et oportet
quod canne per quas trasfertur aqua /
sint de terra cocta et simul commisse et
circa eas murari cum terra gretis / Et
post[e]a super eas murari oportet cum
lateribus calcena cum rena fluminis mista
cum aqua dulci / Et quando vultis quod
illa aqua asciendat implete totam cannam
oribus obturatis prius / et postea deser-
rentur ora ad unam vicem / Et sic prout
habes in designo aqua trahitur de pelago
aut lacu de planitie / Et tunc versatur
aqua per aliam partem decliveam. Et de
inbuendo dictas cannas / modum desig-
num indicaverit in libro leonis ex manu
prefati Ser Mariani Iacobi a fo[lio] 70
ubi ipparet de molendino propter mercur-
ium volvente.

The machine that appears in back, with
a pipe rising over a mountain, is very
admirable, since by [use of] its valve
system water always rises and descends
by inherent motion. In the first place
let the descent be greater than the ascent
by one-fourth. Let the water-carrying
pipes be made of terracotta and connected together, and let clay be walled
up around them. Then let masonry be
placed on them with bricks, lime, and
river sand, mixed with fresh water.
When you want the water to ascend,
fill the entire pipe, having first closed
the mouthpieces. Then let the apertures
be unsealed, both at the same time. Let
it be as you have it in the drawing.
Water is drawn from the pond or lake
in the plain, and then it pours from the
other and downwardly sloping part. As
to the starting of said pipes, the method
is shown by a drawing in the Book of
the Lion, fol. 70 [II 73v], done by the
aforesaid *Ser* Marianus Jacobi, where
also appears the mill driven by mercury [II 74r].

THE DRAWINGS AND TEXTS
OF BOOK IV AND THEIR
HISTORIC POSITION

5 IV 58r, 59r, 60r, Surveying. (Plates 75, 77, 79). The recto of folio 58 has at the top the legend *Incipit quartus liber de edifitiis cotidianis*, Here Begins Book IV, on Commonplace Devices. Alterations of the line were begun but not finished; their meaning is enigmatic. The versos of the present sheets have a total of 45 lines of text with vignettes of a stag, a porcupine, and a bear cub (Plates 76, 78, 80).[116]

As on the first folios of Book III, Taccola again begins with elementary demonstrations of *mensura*. Except for the use of a magnetic compass, IV 60, the methods shown here are modest in comparison with those shown for instance in the *Dioptrica* of Heron, or the *Tractatus astrarii* of Jacopo and Giovanni de' Dondi, or the *Tretis of the astrolabie* by Taccola's older contemporary, Geof-

frey Chaucer, all prior to 1400.[117] Nevertheless Taccola's simple geometry (developed later in such copybooks as Cod. S IV 5 and Cod. Lat. 2941 and in the *Ludi matematici* of Alberti) may have some historic significance in that it taught the elements of this science to builders who might not have known even these beginnings. In a somewhat similar spirit, although in very different style, Kyeser recommended a study of astrology to his readers, leading them in effect to instruments similar to those used here.

IV 61r. Tunnel or Mine. (Plate 81). No text. Vignette of a *Girafa* (Plate 82) on verso.

The drawing on the recto is obviously unfinished. Its finished form occurs in the *De machinis,* Lat. 28800, 33v, 34v. It also occurs in the copybooks based on that treatise,[118] as well as some of those based on *De ingeneis,* the latter copies being likely to reflect some of the lost

---

[116] Folio 58v also has a triangular diagram, in a circle, crudely attempting to define the concept of *dependentia*. Copies of the *recto* drawings are, respectively, in Palat. 767, pp. 23, 26, 22, and Add. Ms 34113, 42v, 43r, 43v. From here on, the copyists reflect the unfinished nature of Book IV. For instance, IV 61r is in Palat. 767, p. 27, and in Ital. Z 86, 44r, but not in Add. Ms 34113. Copies of IV 62 to 68 are in Palat. 767, in Ital. Z 86, and in Add. Ms 34113, as are IV 70 to 75 (except that 71v–73r is lacking in Palat. 767 and in Ital. Z 86). No copyist shows the figure drawings of IV 69r, 76r. A few drawings of Book IV appear in S IV 5.

[117] About the decline of surveying and geometry in Roman times, see Daremberg-Saglio, *Dictionnaire,* s.v. *agrimensor,* and W. H. Stahl, *Roman Science,* Madison, 1962, pp. 85 f, 173, 239. Not enough is known about the gradual rediscovery and development of Hellenistic methods, but see D. Chilton, "Land Measurement in the Sixteenth Century," *Transactions, Newcomen Society,* XXXI, 1961, pp. 111–129, especially pp. 122 ff. Also see Heron, *Opera,* III, *Rationes dimetiendi et commentatio dioptrica,* rec. H. Schoene, in ed. W. Schmidt; S. A. Bedini and F. R. Maddison, "Mechanical Universe. The Astrarium of Giovanni de' Dondi," *Transactions, American Philosophical Society,* n.s., LVI, 5, 1966; R. T. Gunther, *Chaucer and Messahalla on the Astrolabe,* Oxford, 1929. The use of a magnetic compass, underground, was not known in Sienese *butino* practice, according to Bargagli-Petrucci, *Le fonti,* I, p. 41.

[118] Spencer Ms 136, 24v, 25v; Lat. 7239, 51v, 52v; Lat. 2941, 110v.

originals. The *De machinis* and some of the copybooks also include descriptions that refer to the use of gunpowder, in a walled-off section of a mine reached by a fuse, to blow up a fortress. Exact copies are given in most of the copybooks mentioned here, but a slightly modified copy appears in Urb. Lat. 1757, 44v (followed by a blank page).

Taccola may have hesitated to show this method to Sigismund. However, it does not follow that Taccola had invented the method, as is often assumed. It follows even less that Francesco di Giorgio invented it, or perfected it, as has been assumed in the literature beginning with Francesco's younger contemporary, Vannoccio Biringuccio, *Pirotechnia*, 158r. In reality, the assumptions and conclusions of Biringuccio and his followers are wrong.[119] The use of fire in mines goes back to Vegetius; the use of gunpowder in mines goes back at least to 1403. The latter proposal apparently was transmitted to Taccola by Brunelleschi. In a war of Florence against Pisa shortly before the old harbor town was acquired by its envious neighbors, Domenico di Matteo proposed to blow up a city wall by a buried charge of gunpowder. A traitor from Pisa designated a suitable spot. The attempt failed only because the defenders found out in time. Later Domenico was associated with Brunelleschi in another siege effort by the Florentines, this one before Lucca, in the war that Sigismund ended at the time Taccola wrote *De ingeneis*. It is not recorded how often or on what exact dates Taccola saw Brunelleschi, but Brunelleschi's influence on Taccola's ideas about siege methods and related tactics can be found at various places in Mariano's work, including pages of the Sequel to *De ingeneis*, where Brunelleschi is mentioned (S 107r, 107v) and on closely related pages of *De machinis* (Spencer Ms 136, 10v; Lat. 7239, 11v).

| IV 62r. Post Puller. 62v, 63r. Treasure-Hunting and Treasure Hunters Tools (Plates 83, 84, 85, 86). 11 and 19 lines of text on the versos (Plates 84 and 86). The first of these chapters has no vignette. Folio 63 verso has a cat vignette (Plate 86).

Taccola returns here to the subject of harbor constructions and repairs, and the search for ancient treasure previously illustrated in the *capitula* and at III 44. Although elaborately drawn, 63r is not quite complete according to the text. It does not show the last modification of the lantern, with letter "a" Adjustable pincers are shown here, which are known from Theophilus.[120]

| IV 64r. Windmill. 65r. Crank-Operated Mill. 66r. Second Drawing of a Floating

119 A. Pantanelli, *Di Francesco di Giorgio Martini*, Siena, 1870, pp. 36, 61, 93. The principal documents are published by Gaye, *Carteggio*, I, pp. 86, 87, 126. While they are overlooked, a series of nonexistent sources and invalid conclusions are presented in the work of P. A. Guglielmotti, *Storia della marina pontifica*, Florence, 1871, II, pp. 169 ff; V, pp. 22 ff; and unfortunately the pseudoerudition of this work is proliferated in E. Rocchi, *Le origini della fortificazione moderna*, Rome, 1894, pp. 4, 17 ff, and in L. A. Maggiorotti, *Architetti e architettura militari*, II, Rome, 1937 (also 1966), pp. 13, 19, 139. These writers present much rhetoric but no significant facts about events of the time of Popes Nicholas V and Calixtus III, 1447–1458. Their views in turn reappear in historic summaries such as C. Chledowski, *Siena*, II, Berlin, 1905, pp. 187f.

120 Also see Theophilus, *Diversarum artium schedula*, ed. Smith, 1963, p. 9.

Mill. (Plates 87, 89, 91). Folio 64 has 11 lines of text and a fox vignette on the verso (Plate 88). Folios 65 and 66 have no text but have, respectively, a monkey and an elephant as vignettes (Plates 90, 92).

Windmills were a matter of much fascination and of considerable actual use in the late Middle Ages. Various forms are shown by the Hussite author (Lat. 197, Hussite Anonymus, 19r, 19v, 47r) and are diagrammatically redrawn in the reproduction used by T. Beck. Taccola's illustration is strange; it is not clear whether it is based on ignorance of details, or on secrecy, or both.

In any event, Taccola proposes to use the gear-shifting clutch of III 36r again at this point. The proposal seems based on an overestimation of Brunelleschi's reversible drive. A motion-reversing clutch unit was an excellent means to avoid "waste of time" when animals were the power source and when reversals of motion were frequently needed, but in a windmill drive the situation is very different. Since the wind changes direction, whether it should or not, the reversing clutch, applied here, causes more trouble, looseness, and inefficiency than real advantage. Much better means were known to compensate for shifting winds in the operation of windmills with horizontal sails, as schematically shown here; these means consisted of movable shutters. It seems probable that Taccola knew nothing of them. His terms are so uncertain as to suggest that he had only vague, secondhand knowledge of such mills. This, of course, was a poor basis for the attempt to introduce

a Brunelleschian improvement into the construction.

Folio 65 shows a much more plausible device, and one which Taccola presents with clarity and with technical as well as artistic style. To what extent the early Quattrocento knew the crankshaft, crankrod, and flywheel is a matter of controversy, but it is clear that the Hussite author, among others, knew all three.[121] Taccola knew the first two. Nothing, it seems to us, was added to this part of mechanics by Francesco di Giorgio, although he is often cited as an important contributor. Whether the sources now known are sufficient to demonstrate the migration of an idea along some definite path is another question.

| IV 66v–67r, 67v–68r. Trebuchets. (Plates 92–93, 94–95). No text. Vignettes of a griffin and a weasel (?) on 67v, 68v (Plates 94, 96).

Taccola shows two slightly different types of trebuchet. The winding-up mechanism, a horizontal-shaft winch, is shown on 67v. Its placement, but not its form, is similar to the vertical-shaft capstan on 66v connected to the *manganum* on the next page, 67r. Taccola seems interested in the motion of the trebuchet. He refers to this motion when he shows similar although more static machines, such as the crane at III 52. There he refers to the device inscribed *manganum*, which is the title of 67r.

Earlier and better illustrations of trebuchets can be found, for instance in a

[121] For the Hussite part of Lat. 197, 18r, 21v, 42r, 43r, see Beck, *Beiträge*, figs. 314, 315, 317, and Berthelot (1891), fig. 22.

rather famous part of Kyeser's book.[122]
Sigismund, as mentioned, had known
Kyeser. It is conceivable that the old
warrior Sigismund influenced the course
of Taccola's progress in developing the
book. The preface, referring to the questions asked by the king, and the possible admonishment in the event that
others knew more, would be reconcilable
with such a hypothesis. However, it
could be nothing but a hypothesis.

IV 69r. Dedication. Picture of St. Dorothy and the Christ Child. (Plate 97).
29 short lines of lettered writing, with a
marginal correction. The verso is blank.

The saint is shown here, near the end
although not at the end of the book. The
text, composed as if spoken by St. Dorothy (*Sancta Oradea*), contains Mariano's
petition to be made *familiaris* of the
Emperor-elect, and it also speaks of
codices wherein Taccola could describe
and illustrate the history of the kings
of Hungary up to Sigismund's own
time. He may have visualized a work
like that of his Sienese contemporary,
Niccolò di Giovanni Francesco Ventura,
who wrote and illustrated *La sconfita di
Montaperti*.[123] It seems pertinent to quote
what Sigismund's chancellor, the humanist Vergerio the Elder, wrote about this
time, ca. 1433: "You have ordered me, o
Sigismund, most humane emperor, to
translate Arrian's history [of Alexander the Great and his successors] into
Latin. . . . The plan is much to be ad-

mired. . . ."[124] It seems possible that the
plan became known to Taccola and his
humanist friends during the weeks between Sigismund's arrival in Siena and
the formulation of the present text. The
book connecting Sigismund with Alexander was then written by Vergerio, but
it is not known whether Taccola had occasion to make miniatures for it.

IV 70r, 71r. Special Cheval de Frise.
Mechanized Cheval de Frise. (Plates
99, 101). No text. Camel vignette with
the year "1432" on 70v. (Plate 100). No
vignette for 71.

These devices are variants of drawings
that occurred mainly in the *capitula* and
at some other places in Books I and II.
Together with the preceding *tribucae*,
they reappear in the later *De machinis*,
although in slightly modified forms.

IV 71v–73r. Automatically Propelled
Boat. (Plate 102). No vignette and no
text. However, the illustration was to
be described. Letters "c," "d," and "f"
are shown on the main drawing and on
the smaller diagram on the right side.
In all other cases such letters are cited
by Taccola in a descriptive text. That the
text is lacking here is one of the indications that Palat. 766, at least toward its
end, becomes only the draft of a more
finished work then extant or planned.

The device shown here is operated by
the illustrated waterwheels, which are
turned by the flowing river. The wheels
turn a winch, winding up a rope, which
causes the boat to wander upstream
against the current. The idea was similar

---

[122] Kyeser, *Bellifortis*, 30r, in Berthelot (1900),
p. 312.

[123] See B. Degenhart and A. Schmitt, *Corpus
der italienischen Zeichnungen, 1300–1450*, IV,
Berlin, 1968, pp. 321 f and Pls. 239–240.

[124] P. P. Vergerio, *Epistolario*, ed. L. Smith
(*Fonti per la Storia d'Italia*, LXXIV), Rome,
1934, pp. 379 ff; Voigt, *Wiederbelebung*, p. 273.

to proposals known to Kyeser (*Bellifortis,* fol. 134r) and may have been proposed in earlier times. Certainly others repeated it or believed that they had invented it themselves. Among them was Bishop Verantius (1615) as well as the American frontiersman Rumsey, sponsored by George Washington (1784).[125] When actually tried, the device never performed satisfactorily on rivers.

❙ IV 73v. *Finis.* (Plate 103). No vignette. Taccola wrote Books III and IV but had come to hesitate about a number of elements: the *incipit* to Book IV, some of the book's contents, such as the unfinished folio 61, and the texts for machine drawings, beginning with folio 65, which were not written. At this point he proposes to finish a "Part III." This was not his last word; he inserted further technical drawings and another final picture (IV 76r). On 13 January 1433 [1432], he may have intended to finish his manuscript and to make it consist of a single part or book. However, he did not then actually complete it and did not, so far as Palat. 766 shows, make it a one-book unit.

❙ IV 74r, 74v, 75r. Roofbeam Joints. (Plates 104–106). No text or vignette on these folios or on 75v (Plate 107).

The drawings on these three sheets, like the corresponding ones in Book I (37r, 48r) and the Sequel (S 120v, 121r) are likely to be copied from some carpenter's notebook. They in turn are copied faithfully in the copybooks.

It appears that folio 75v was at one time intended for the drawing of some wind-driven mechanism, because a wind symbol, a winged head with puffed cheeks, on 76r (Plate 108), blows in a direction toward this page. Actually the page was then merely used to show a burning mirror reflecting the sun to ignite a candle or the like. Perhaps the wind symbol is intended to extinguish the flame, although it does not blow down toward the flame. There is no text to elucidate this drawing.

❙ IV 76r. St. George Slaying the Dragon. (Plate 108). No text or vignette. The verso is blank except for a note added by a later hand (Plate 109).

A princess in chains, wearing a diadem somewhat similar to that of the Emperor-elect on the frontispiece, 27v, is liberated from the dragon by the saint with halo, sword, and shield. This drawing, on the last page, seems calculated to refer back to the beginning of the "little book," *libelli,* to use Taccola's word. The subject of this drawing, a warrior-saint, may have had special appeal to the proposed sponsor whose portrait as warrior appears on the first folio.

---

[125] Faustus Verantius, *Machinae novae,* Venice, n.d. [1615], p. 40. Concerning Rumsey, see J. T. Flexner, *Steamboats Come True,* New York, 1944, pp. 68, 384, where an English reference of 1663 is mentioned. See also Beck, *Beiträge,* p. 289 and fig. 351, for a parallel drawing from Book II.

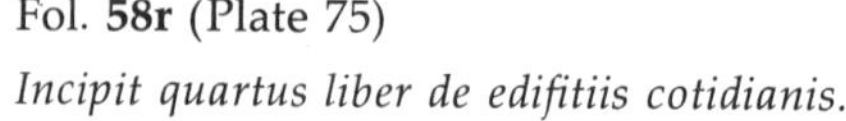

75. Surveying a water course and
Incipit of Book IV **(IV, 58r)**

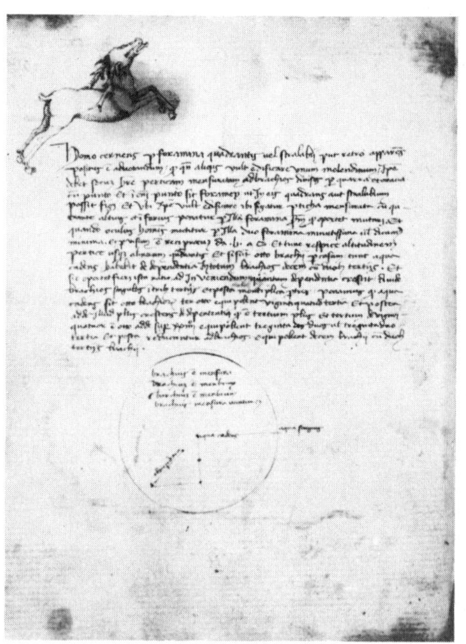

76. Text relating to surveying **(IV, 58v).**

Fol. **58r** (Plate 75)

Incipit quartus liber de edifitiis cotidianis.

Fol. **58v** (Plate 76)

Homo cernens p[er] foramina quadrantis
vel stralabii prout retro apparens positus
est ad notandum / quod quando aliq[u]is
vult edificare unum molendinum / ipse
debet secum habere perticam mensuratam
ad brachios divisos per quarra et ottava
cum punto Et in omni punto sit foramen /
ut in eis quadrans aut stralabium possit
figi / Et ubi ipse vult edificare ibi figatur
p[er]tica mensurata / cum qu[a]trante
altius et inferius ponatur per illa foramina
secundum quod oportet mutari. Et quando
oculus hominis mictitur per illa duo for-
amina minutissima vel dicamus minima.
Et per visum est reciprocus da .b. a / c.
Et tunc respice altitudinem pertice usque
a lineam quatrantis Et si sunt otto brachii
per casum / tunc aqua cadens / habebit
de dependentia in totum brachios decem
cum duobus tertiis. Et sic oportet fieri ista
ratio / ad inveniendum quantum depen-
dentia exestit / Divide brachios singulos
in tribus tertiis Et post[e]a montiplica per
tres / ponamus quod aqua cadens sit otto
brachiorum ter otto equipolent viginti
quatuor tertia / Et postea adde illud plus
crescens de dependentia quod est tertium
plus Et tert[i]um de viginti quatuor est
otto adde super XXIIII° equipollent tre-
ginta duos vel treginta duo tertia / Et*
post[e]a / reducantur ad brachios ./ Equi-
pollent decem brachii cum duobus tertiis
brachii.

* Three letters are canceled after *tregintu*.

A man who looks through the sight
holes of a quadrant or astrolabe is shown
in back. Be it noted that when anyone
wlshes to build a mill, he must have
with him a rod measured in *brachia*
[1 BR. = c.2 feet], divided into fourths
and eighths by marks, with an aperture
at every mark so that the quadrant or
astrolabe can be fixed therein. Where he
wishes to build, there he must fix the
measured rod with the quadrant. He
must raise and lower the quadrant in
said apertures, depending on how it
must be adjusted. When the eye of the
man is put to said pair of sight holes,
or as we say, peepholes, and when it
coincides with sight from "b" to "c,"
then read the height of the rod to the
line of the quadrant. If it is, for instance,
eight *brachia*, than the water descending
should have a total incline of ten and
two-thirds *brachia*. Let the calculation
of the incline be made as follows: divide the *brachia* into three thirds and
then multiply by three. Let us, for instance, assume that the descending water
is eight *brachia* [high], three times eight
equals twenty-four, to express it in
thirds. Then, as increment for the incline, add so much more as is one-third,
and [this] one-third of twenty-four is
eight. Add [this] to the 24 [thirds]. It
equals thirty-two, that is, thirty two
thirds. Then reduce it to *brachia*—it
equals ten and two-thirds *brachia*.

77. Surveying a mountain **(IV, 59r).**

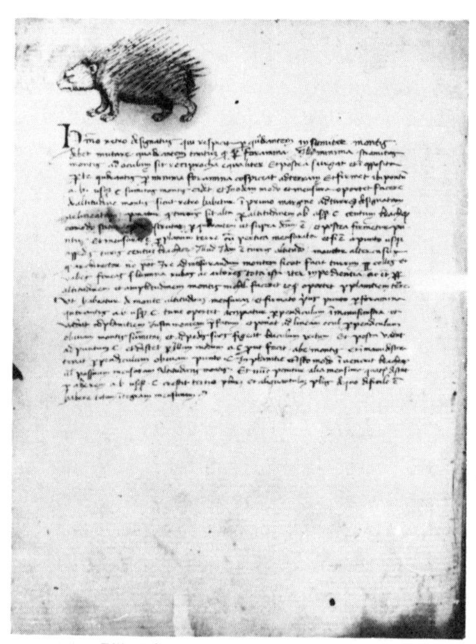

78. Text relating to surveying **(IV, 59v).**

Fol. **59v** (Plate 78)

Homo retro designatus qui respicit per quadrantem in sumitate montis debet mutare quadrantem tantum quod per foramina illius minima / estremitas montis ad oculum sit reciprocha equaliter / Et postea surgat et in opposita parte quadrantis per minima foramina cospiceat ad terram Et firmet ibi pontum a .b. usque .c. sumitas montis cadet / et in [e]odem modo et mensura oportet facere de altitudine montis sicut retro habetur in primo margine ad turrem designatam ac lineatam / Ponatur quod turris sit alta per altitudinem a b usque c centum brachiorum com[m]odo scitur / oportet scrutari per quadrantem ut supra dectum est / et postea firmetur puntus Et mensurare per planum terre cum pertica mensurata et si est a punto usque appedes tur[r]is centum brachiorum illud

The man designed in back, who looks through a quadrant at the peak of a mountain, must change [the setting of] the quadrant until the highest point of the mountain is equally in line with the eye, through the peepholes of the instrument. Then let him rise. From the opposite side of the quadrant let him look through the peepholes to the ground, and make a mark where [the line] from "b," the top of the mountain, strikes [the ground, at] "c." Let him proceed in the same way and measure, as to the height of the mountain, as is done in back on the first page [III 27r], where the tower is designed and delineated. Assuming that the tower is two hundred feet high according to the height of [line] "b"–"c," this is readily deter-

idem est turris altitudo / movetur alter
casus quod architector non potest ire ad
mensurandum montem sicut facit turrim /
propter colles et valles foveas flumina
rubus ac arbores tota ista iter[i] inpodien-
tia / ac etiam propter altitudinem et am-
plitudinem montis nichil faceret eo quod
oportet per planitiem terre ut habeatur de
monte altitudinis mensuram / et firmato
prius punto per foramina qu[a]trantis a b.
usque .c. tunc oportet accipiatur perpen-
diculum in manu sinistra et vadat ad plan-
itiem iusta montem prefatum / et ponat ad
lineam oculi perpendiculum obviam montis
summitati et ad pedes suos figeat baculum
rectum / Et post[e]a vadat ad puntum .c.
et distet per illim modum a c prout fecit /
abc / montis / et in manu destera tenat
perpendiculum obviam punto .c. in planitie
et in [i]sto modo invenirit brachiorum vel
passuum mensuram altitudinis montis. Et
nunc ponitur alia mensura qua[n]tum distat
per aherem a .b usque .c. crescit tertio
pluri[s] Et aliquantulum plus de quo dificile
est habere totam integram mensuram etc

mined. It must be viewed through the
quadrant as said above; the point must
then be marked; and [you must then]
measure on the flat ground with a measuring rod. If it is two hundred feet from
the point to the foot of the tower, the
same is the height of the tower [when
the quadrant is set as shown]. Let another case be assumed, that the architect
cannot proceed to the measurement of
the mountain as he does with the tower,
because of hills and valleys, depressions,
rivers, bushes, and trees—all that which
obstructs his path—and also because of
the height and breadth of the mountain,
that he can do nothing in such way as
he would on flat ground, in obtaining
a measure of the height of the mountain.
[In that case,] when he has first marked
the point [where the line to the ground
ends,] from "b" to "c," let him take a
plumb line in his left hand; let him go
to a flat spot near the aforesaid mountain; set the plumb line opposite the
peak of the mountain as seen with his
eye; fix a staff upright at his feet; then
walk to point "c," marked [at the end
of line] "b"–"c" from the mountain; and
let him hold in his right hand a plumb
line opposite point "c" on the flat spot—
thereby he will find a measure of the
height of the mountain in feet or paces.
Then let another measure be assumed
for the distance from "b" to "c" by air.
It is increased by one-third, and a little
more, for which it is difficult to obtain
a complete and perfect measure.

79. Surveying for tunneling **(IV, 60r).**

80. Text relating to surveying (IV, 60v).

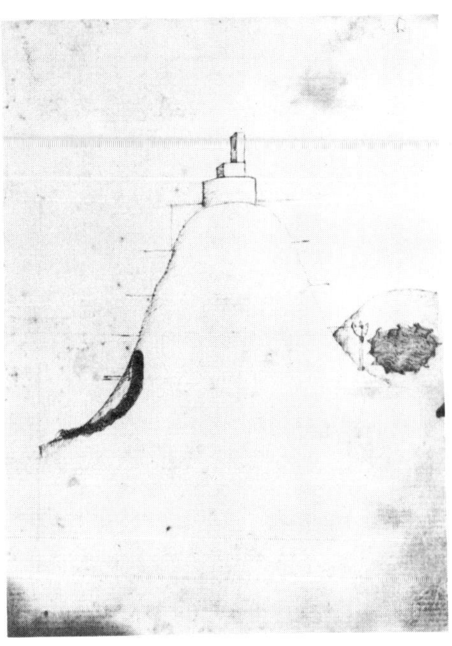

81. Tunnel or mine (IV, 61r).

Fol. **60v** (Plate 80)

Montem qui retro videtur cum tribus sguat-
ris et funibus ac perpendiculis si aliquis
voluerit fovere per medium ab uno latere
ad b alterum / respiciatur in dicto monte
cum mensuris et perpendiculis suis per se
ipsum satis est manifestum. ¶ In medio
montis prout videtur est bussula que fertur
in mare a nautis / est valde bona sub terra
quando foratur mons. Et propter ipsam
bussulam scitur sub terra versum quam
regionem huomo existit / quia propter
calamitam demostratur stella tramontana
eo quod de naturu sui attru[h]it ad se
calamitam per quam illico indicatur stella /
per quam docetur cavatoribus via caverne
vel tunbe [per] quam linea[m] recta[m]
facere possunt ad fovendum.

If someone wishes to dig through the
middle of a mountain, from one side to
the other, "b," as is shown in back with
three level instruments and with ropes
and plumb lines, let surveys be taken on
said mountain with measuring instruments and their plumb lines, [as] is
fairly obvious by itself. ¶ Inside the
mountain, as is seen, [there] is a compass of the kind that is used on ships
at sea. It is very useful underground
when the mountain is being perforated.
By means of the compass one can know
underground in which direction a man
faces. For the needle points to the north
star since it, by its nature, attracts the
needle to itself. Thus the star is readily
indicated, whereby diggers learn the
direction of the hollow or underground
passage, so that they can dig in a straight
line.

82. Giraffe eating the fruit of a date palm
(IV, 61v).

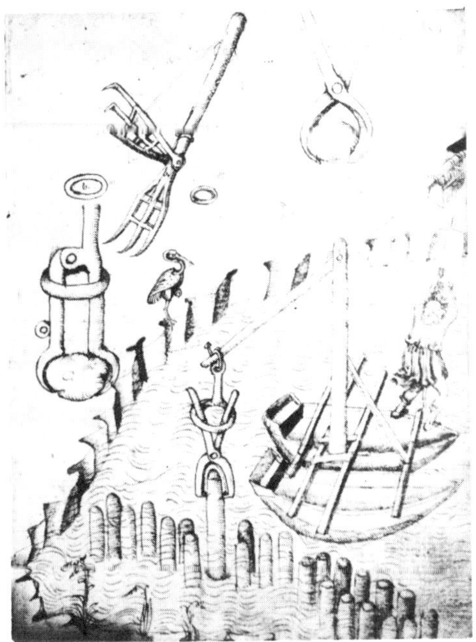

83. Post puller **(IV, 62r).**

Fol. **62v** (Plate 84)

Stagnum retro habens in se multos palos
propter quos non potest navigari / |sive|
per aliquen modum transire / tunc oportet
quod ibi habeantur duo navigia / vel cinbe
coppulate habentes in se stilum rectum
cum (cum) pertica trasversali cum tanaleis
ferri prout videtur in designo / Et si huomo
non potuerit per vim suam extrhare palos
tunc faciat cum argano prout retro ap-
paret quod melius dicitur cum naspo ad
ovem pascentem e[t] de om[n]ibus aliis
tanaleis de super in isto margine desig-
natis / bone sunt ac trhaendum lapides
lignamina et lutum Et de aleis tanaleis
actibilibus add alia negotia apparet in
libro dicti Ser Mariani adlecti in libro*
primo leonis u fo[lio] 20 .—

" The reading *adlecti* (meaning *allegati* ?) is
uncertain, and it seems possible that Mariano
canceled it. However, a canceling line is not
clearly visible.

84. Treasure-hunting tools. Text relating
to the post pullers **(IV, 62v).**

In back: a lake, with many posts in it,
due to which it cannot be navigated or
in any way passed. In such case let two
boats or vessels be provided, which are
joined together and carry a vertical post
with a transverse beam and with iron
crampons, as shown in the design. If a
man cannot extract the posts by his own
strength, let him use the capstan shown
in back, better known as a winch, near
the Pasturing Sheep [III 34r, 34v]. What
concerns all other crampons additionally
designed on this sheet, they are useful
for lifting stones, wood, and mud. Other
crampons for other work appear in the
book of said *Ser* Mariano, aforesaid—
in the First Book, of the Lion, on folio
20 [I 18r, 18v].

85. Treasure hunters **(IV, 63r).**

86. Text relating to the treasure hunters **(IV, 63v).**

Fol. **63v** (Plate 86)

Navigium retro apparens cum duabus
figuris est bonum / cum illis instrumentis
ad coligendum quidquid ceciderit in
mare(m) aut flumina / Et multotiens evenit
propter alititudinem aque non videtur
fundum / Et propter hoc oportet habere
in fundo maris lumen / Et tunc in casu
isto decet habere vasum vel vax[um] ad
i[n]star laterne confectum de terra cocta
habens in summitate manubrium cui funis*
applicari poterit et intus mictatur lumen /
et dectum vax[um] sit totum serratum
nisii in parte inferiori tubum sit apertum
et mictatur in aquam intus ad videndum
rem suam quam ipse huomo rimatur / Et
si est urceus auri vel argienti vel cuiusque
generis existat faciat cum storpionibus.

* In the manuscript appears *summitatem*, with
the last *m* erased.

The ship with two figures which appears
in back is useful [when employed]
with the [illustrated] instruments for
collecting what may have fallen in the
sea or in a river. It often happens that
due to the depth of water the bottom is
not seen. Therefore let a light be provided on the bottom of the sea. For this
purpose, then, it is suitable to provide
a container or receptacle similar to a
lantern, made of terracotta, with a handle
on top where a rope can be attached. Let
a light be put inside and let the receptacle be entirely closed, except that in
the lower part it be [like] an open tube.
Let it be lowered into water so that the
man may see the object for which he
searches. If the object is a pitcher of
gold or silver, or of whatever kind it

*Et si est anelus faciat cum perpendiculo
plunbii et in parte inferiori sit cum pice
navali aliquantulum liquefacta cum seta /
et tunc gravetur† super anelum prefatum
perpendiculum / Et aureum trhatur perpendiculum Et invenerit anelum applicatum
in pice / movitir nu[n]c casus si fundum
in quo est anelus est lutosum et dubitatur
ne intret magis intro propter gravedinem
plunbi / tunc oportet agere cum tanaleis
retro ad piscem / Et movitur alter casus
quod urceus cum storpionibus non habeatur / tunc recurratur retro ad navigium
habens secum forfecem signatam .b. et
super urceum ponatur forfex statim per se
aperiatur et carpit eum / contigit aliud
dubium / si quis non haberet laternam
terrecoctam / in casu isto habeat laternam
bene costrictam lignaminibus bonis a labor-
atam ac cum saxis pendentibus circiter
eam gravantibus ad fundum cum lumine
intus laternam. Et habebit obtentum suum :—*

† Before *super,* the same word appears with a
line through it.

may be, let him operate with pincers
[to retrieve it]. If the object is a ring,
let him operate with a plumb bob of
lead; in its lower part let it be slightly
moistened with naval pitch by means of
a brush; then let the bob sink onto the
aforesaid ring, draw up the bob, and the
ring will be found stuck to the pitch.
Considering the case that the ring is in
muddy ground and that there is doubt
whether it will be driven deeper into it
by the weight of the bob—in this case
it is necessary to work with pincers as
shown in back on the sheet with the
fish [IV 62v]. In the further case that a
jug cannot be grasped by pincers with
teeth, refer back to the sheet with the
ship provided with tongs marked "b"
[IV 62r]. When these tongs are placed
on the pitcher, they open at once, by
themselves, and catch it as they close.
Another question of doubt—when there
is no lantern of terracotta, let a lantern
be well constructed, made of good lumber "a," with stones hanging around it
to weigh it down to the bottom and with
a light inside the lantern; he will have
what he is looking for.

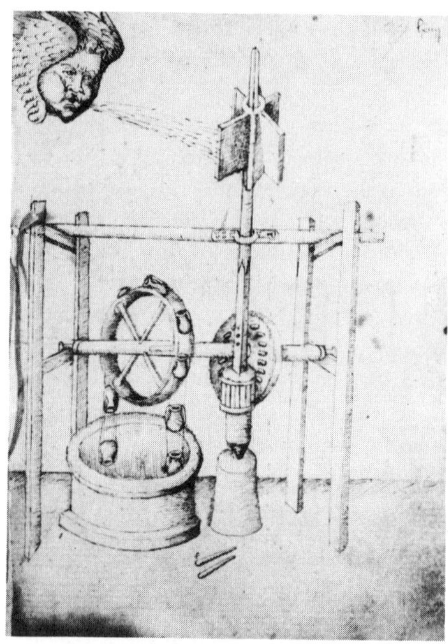

87. Windmill **(IV, 64r)**.

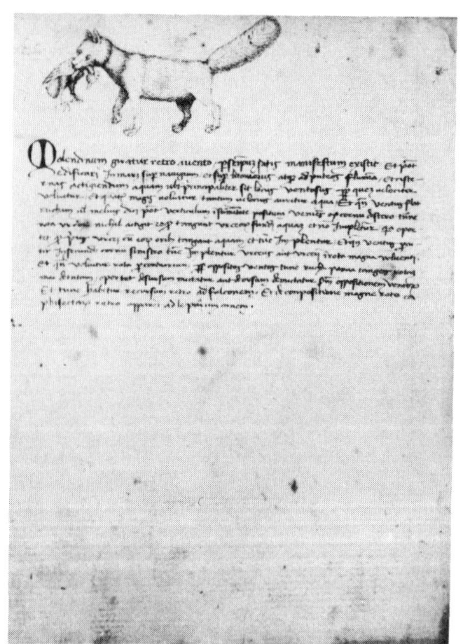

88. Text relating to the windmill **(IV, 64v)**.

89. Crank-operated mill **(IV, 65r)**.

Fol. **64v** (Plate 88)

*Molendinum giratur retro a vento / per se
ipsum satis manifestum existit / Et potest
edificari in mari super navigium / et super
litoracius atque ad puteos flumina / Et
cisternas actigiendum aquam ubi princi-
paliter sit locus ventosus / propter quem
velociter volvatur / Et quantum magis
volvitur / tantum velocius [h]auritur aqua /
Et quando ventus flat rucham vel melius
dici potest verticulum in summitate posi-
tum veniens ex cornu destero / tunc rota
urceorum nichil actigit / eo quod tangunt
urceorum fundi aquam et non impleantur /
quia oportet quod prius urcei cum eorum
oribus tangant aquam / et tunc implentur / Et quando ventus percutit in secundo
cornu sinistro tunc implentur urceus aut
urceii in rota magna volventi. Et quando
volvitur rota per contrarium propter op-
positos ventos tunc rucha parva tangens
rotam* de[n]tatam / oportet desursum mictatur aut deorsum demictatur secundum
oppositionem ventorum Et tunc habetur
recursum retro ad falconem. Et de con-
positione magne rote caphifectarum ratro
apparet ad lepuarum canem.*

* *tangens* is misspelled *tangeis*. After *rotam*
appears *ma*, with a line through it.

How the millwork, in back, is turned
by wind is obvious from the drawing.
It can be built on seagoing ships and
on land, for obtaining water from wells,
rivers or cisterns, wherever a windy
place is found, where it revolves quickly.
The more it revolves the more quickly
is water raised. When wind blows from
the right against the upper wheel, the
bucket wheel achieves nothing, since
the bottoms of the buckets then hit the
water and they are not filled. It is necessary that the buckets hit water with the
open ends first, then they are filled. Second, when wind comes from the left,
then the bucket or buckets of the great
revolving wheel are filled. If the wheel
reverses, due to changing winds, let the
small pinion meshing with the large
cogwheel be moved up or down, according to changing winds. For this, refer to
[the sheet of] the falcon, in back [III 38].
For the construction of the great bucket
wheel, see the sheet of the hare-hunting
dog, in back [III 39].

90. Walking monkey **(IV, 65v)**.

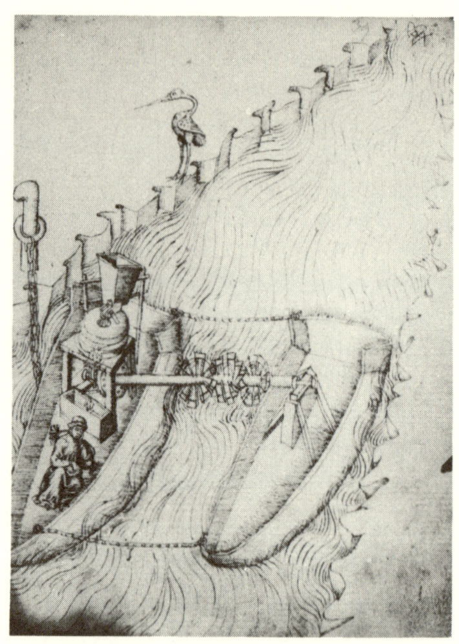

91. Floating mill **(IV, 66r)**.

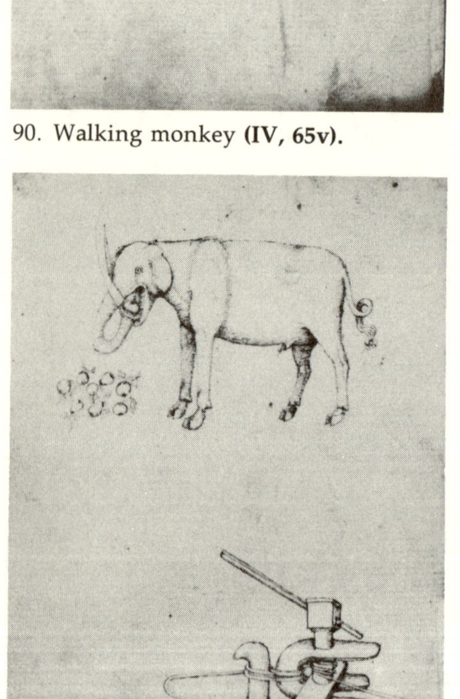

92. Elephant **(IV, 66v)**.

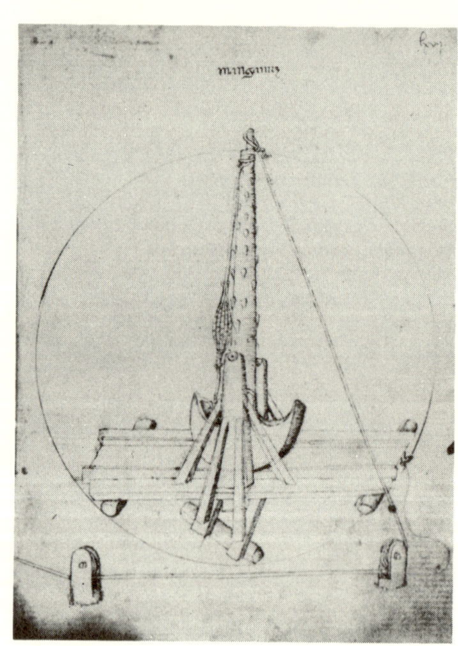

93. Trebuchet **(IV, 66v–67r)**.

94. Griffin **(IV, 67v)**.

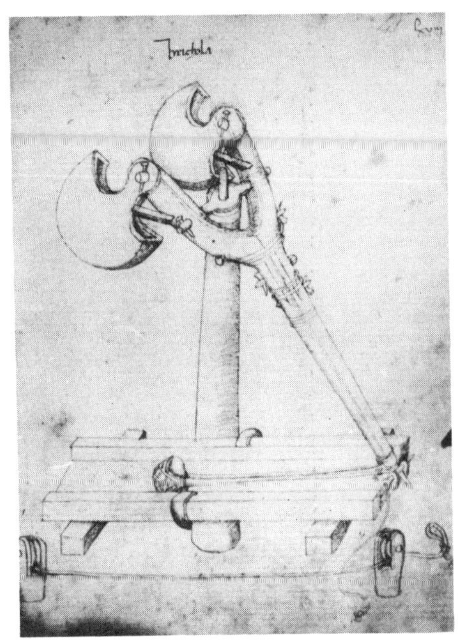

95. Trebuchet (variant) **(IV, 67v–68r)**.

96. Weasel **(IV, 68v)**.

97. Dedication. St Dorothy and the
Christ Child **(IV, 69r)**.

Fol. **69r** (Plate 97)

* *omnia facta sunt* is inserted on the inner
margin.

¶ For you, Sigismund, most exalted
prince, ever venerable King of the Romans, I beseech almighty God and the
word of the Father, the creator of all,
that He may support you in the faith,
in long-lasting health and in His love,
that you may fight against our enemies
and that you may obtain victory over
them, receive the imperial crown of the
entire world, and, on the day of achievement, be led to eternal life, amen. ¶ To
your sacred majesty I always recommend
Ser Marianus Jacobi, secretary of the
Domus Sapientiae in Siena, who composed this little book and drew my picture. May you be pleased to accept him
as one of your family and court, and by
the authority of your power to make
him master of machines for waterworks.
It is his intention to dwell in your Hungarian regions and there to finish his
days; to attend to all the devices for
waterworks; in books to describe all
things done and achieved by you, kings
of Hungary, and your antecedents, according to his ability to compile [what
is known of] every place; and in said
books in margins of the pages to design
and illustrate stories.
ST. DOROTHY.

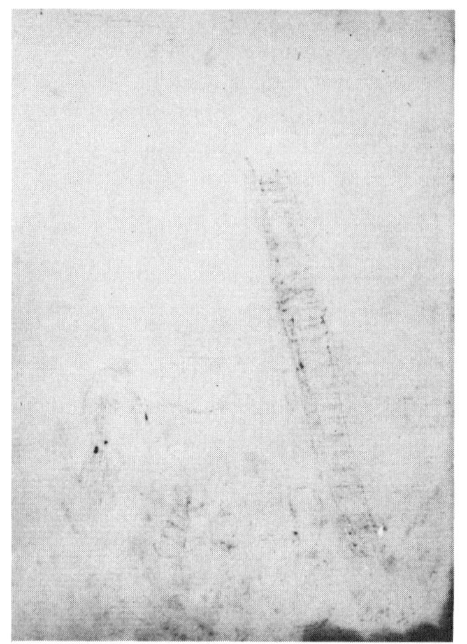

98. **(IV, 69v).**

99. Special Cheval de Frise **(IV, 70r).**

100. Camel. The year, 1432. **(IV, 70v).**

101. Mechanized Cheval de Frise
(IV, 71r).

102. Automatically propelled boat **(IV, 71v–73r).**

103. *Finis* **(IV, 73v).**

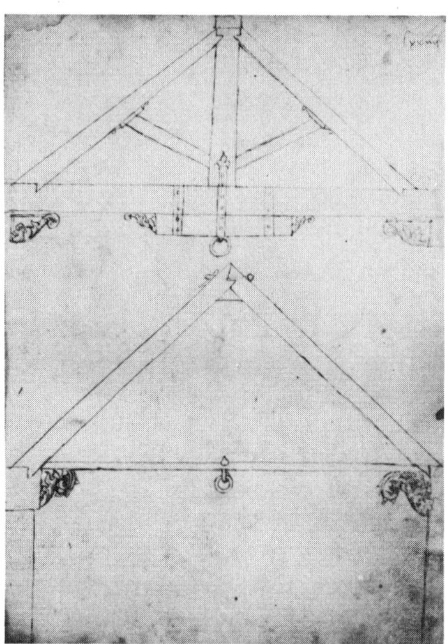

104. Roofbeam joints **(IV, 74r).**

Fol. **73v** (Plate 103)

Feliciter finit tertia pars libelli de edifitiis ac ingeneis completa in Domo Sapientie civitatis Senis in anno domini millesimo CCCCXXXII die xiii mensis Ianuuarii. dum Senenses et Florentini malam viciniam peragebant. :-

Thus propitiously endeth the Third Part of the little Book of Devices and Engines, completed in the Domus Sapientiae in the city of Siena on 13 January 1432 [1433], while the Sienese and the Florentines acted as bad neighbors.

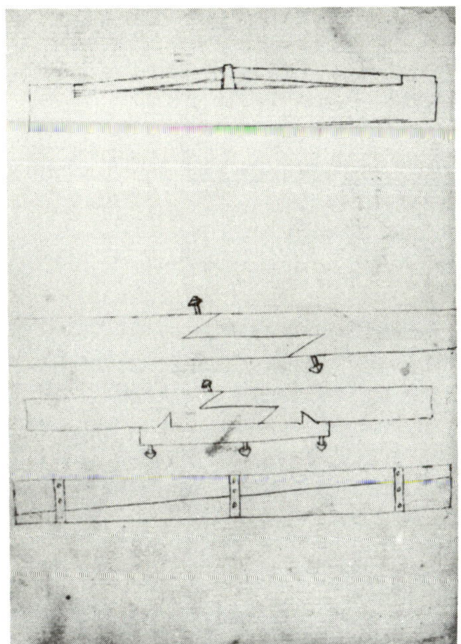

105. Roofbeam joints (variants) **(IV, 74v).**

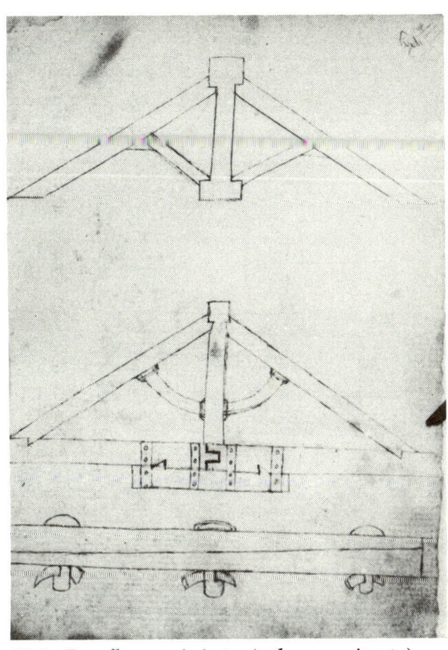

106. Roofbeam joints (other variants) **(IV, 75r).**

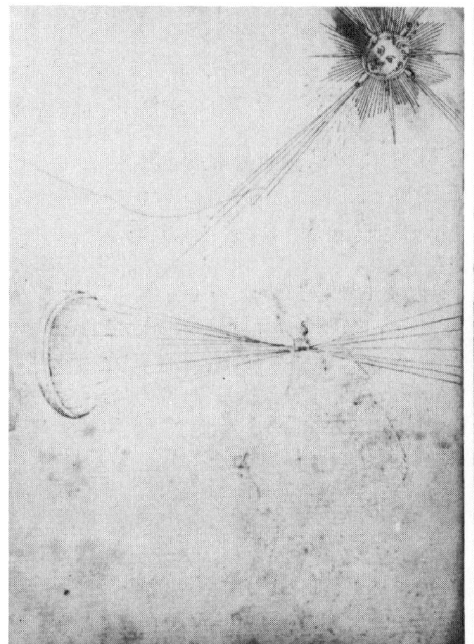

107. Burning mirror reflecting the sun
(IV, 75v).

108. St. George slaying the dragon
(IV, 76r).

109. Notes (upside-down) by a later
writer **(IV, 76v).**

III

MARIANO TACCOLA
AS ENGINEER

6 The technical concepts recorded
by Taccola are rational. They deal with
utilitarian devices, and at many points
they are developed to improve such
devices for their utilitarian purposes.
The viewpoint is very different from that
of the philosophers and scholastics of
any period who seek a permanent system of the world. Taccola seeks progress.

He formulates this aim most clearly in
his ultimate work, *De machinis,* where he
repeatedly inserts aphorisms and verses
praising the value of *ingenium* and *sensus,* the sources of technical progress.
He describes them as the origin of civic
improvement; where he illustrates one
of the basic hoist mechanisms (Lat.
28800, 18r), he writes, *Plus valet ingenium
quam bubalorum vires.* More is done by
genius than by the power of oxen. He
also describes ingenuity as the origin
of improvement and success in the arts
of warfare. Where, on folio 57v, he illustrates an armed and mounted soldier
with various paraphernalia of military
accoutrement, he either composes or
quotes from one of his contemporaries:
*Hic vigor ac sensus equali munera lanze
pensant / alcidem vigor armat sensus
ulixem.* Here (in war) vigor and sense
are, for the work of the lance, of equal
weight. Heracles fights with vigor,
Ulysses with sense.

He is less poetic in *De ingeneis,* but his
viewpoint is clear. He wants engines of
greater speed, power, efficiency, and
economy. He assumes that masters want
to prevent "waste of time" in raising
"any large load." He hopes to devise
mill systems for grinding corn with
less need for falling water, and foundation methods for building harbors and
bridges more rapidly. With Brunelleschi
he hopes to harness energies more effectively where reversals of motion are
needed. He also wants more durable
cannons, more forceful powder, more
accurate aiming, more rapid transporta-
tion, and mines more effectively placed.
The progress-oriented attitude prevails
even when he occasionally returns to the
more stationary system of the microcosm
and its magic qualities; he shows it
with accurate tools for measuring its
proportions and measures them more accurately than others had done.

However, he is only at the first beginnings of a hard and steep climb. Engineering knowledge had declined badly
in former centuries. The constructions
known to him fail to include many of
the more refined devices developed in
antiquity. He does not mention roads.
His surveys are crude. In the hydraulics
that he has inherited, such a famous
thing as the Archimedean Spiral (I 38v)
is totally misconceived, although Vitruvius (X vi) had stated its requirements
clearly. Taccola's text reflects many ideas
for the construction of pumps, pump
systems, mills, hoists, and other devices
but does not reflect knowledge of the
materials that antiquity found suitable,
and which were again found suitable
shortly after Mariano. Among these

materials were bronze for machine parts in need of machining, or truly durable cements for buildings. His automated mill (III 33v–34r), is well ahead of his time; on the other hand, his method of using measurements in surveying a mill race (IV 58v) appears impoverished when compared with some of those used in antiquity.

Ancient methods of using measurements were not absolutely lost but were unknown to the geometers and surveyors of the Middle Ages, and probably even to those of imperial Rome. Two centuries before Taccola, the famous Leonardo of Pisa (Fibonacci) rediscovered some parts of Greek mathematics, probably in Byzantine and Arabic books, and promptly restarted their development. He became famous enough to appear before an emperor, as Taccola later did.[1] However, his beginnings were insufficient to convey any developed mathematical knowledge to those who could have taught Taccola, and Taccola himself shows no awareness whatever of the knowledge developed by Fibonacci in nearby Pisa. Taccola used only such methods of dealing with measurements as were available to the most modest of Roman agrimensors. These methods lag substantially behind those of Hellenistic times, and they developed only minutely in ancient Rome and hardly at all in the Middle Ages. In the famous aqueducts of the city of Rome there had been some progress in fixing the slope, from the oldest to the late imperial structures, some of which continued to have lim-

ited use.[2] Taccola seems to know the late Roman practice in this field. In that practice and in corresponding practices of the Middle Ages considered by Taccola, there had been much stagnation.

Then, about Taccola's time, with his help and that of others, this stagnation came to an end. His *De ingeneis* became the starting point for a long line of copy-books. New concepts appeared in both this autograph and some of its copies. Society gave some recognition to its new Archimedes. A man of the people, like Taccola, could appear before the exponents of governmental power or could hold "council" with men like Brunelleschi. Some such "council," if not such secrecy as was proposed by Brunelleschi in his speech to Mariano, was needed. The amount of technical knowledge accumulated and distributed in books and available in the time of Mariano was small. Greek and Hellenistic works were unknown, and their reviews by Roman writers were forgotten, except to such little extent as they were reflected in encyclopedias and abridgments, such as those of Pliny and his medieval followers. Some of these

[1] M. Cantor, *Vorlesungen über Geschichte der Mathematik,* II, Leipzig, 1913, pp. 6, 40 ff.

[2] The texts of the *agrimensores* or *gromatici,* written ca. A.D. 50 to 350, and collected ca. 450 to 550, are in F. Blume et al., *Die Schriften der römischen Feldmesser,* Berlin, 1848–1852. For the dates and some illustrations, see C. Thulin, "Die Handschriften des Corpus Agrimensorum Romanorum," *Abhandlungen, Königlich Preussische Akademie der Wiss., Philos.-hist. Classe,* part II, 1911. For surveys of aqueducts, see the comments of C. Herschel in his edition of Frontinus, *De aquis urbis Romae,* Boston, 1899, pp. 201–216; "Agrimensores" in Pauly-Wissowa, *Real-encyclopaedie,* I, 1894, cols. 894–895. For other details see our notes to IV 58r ff.

works no doubt were available at the Sienese Studio and Sapientia.[3] A brief review of the technical knowledge reflected by the *De ingeneis* will bear this out.

Taccola's drawings show some dim reflections of information apparently derived from the more elementary chapters of Heron's *Pneumatica,* and there are added reflections in one of the copybooks (Add. 34113, 1 ff: *i libro di Filone delle ingiegni ispirituali*) of the earlier Hellenistic work by Philon of Byzantium. Military science was somewhat better known. A review of the manuscripts suggests that Taccola (like Ghiberti) knew something of writers on the art of war—Athenaeus the Poliorcetic, another Alexandrine, and Apollodorus of Damascus. Their works were known at least to some of Taccola's contemporaries and potential advisors, if not to him personally. It is certain that he, as well as his advisors, must have known the Vegetian *Epitoma rei militaris* or one of its medieval translations, such as the second part of the *Livre des faicts d'armes et de chevalerie* by Christine de Pisan, entitled *De l'ordre et manière de combattre et deffendre chasteaulx et villes selon Vegèce. . . .*[4] Taccola illustrates substantially each machine textually described by Vegetius, or Jean de Meung or Christine, except that he knows little of the catapult and ballista and generally substitutes the medieval, weight actuated trebuchet.[5] Other work, including the *Thexaurus* of Guido da Vigevano (1335)[6] and the *Bellifortis* of Konrad Kyeser (ca. 1405) had illustrated such machines with a liberal admixture of fantastic speculation. Taccola, like his contemporary the Hussite Anonymus, shows them with sober realism. Perhaps, in fact, he becomes more sober, realistic, and conventional in his showing of these machines as he progresses from his early Chapters to his *De ingeneis.* Taccola's *capitula* were full of original and speculative variants of the trebuchet; his later work shows the device as it was, with variants limited to minor details of winding-up winches. He had learned the science recorded by Vege-

---

[3] Concerning Sienese libraries and their administration, see Zdekauer, *Studio,* pp. 85 ff, 93.

[4] First anonymously printed, *L'art de chevalerie selon Vegece, traicté de la manière que les princes doivent tenir au faict de leurs guerres et batailles,* Paris, 1488. It was then edited by W. Caxton, *A Book of Christine of Pyse drawn out of Vegetius de re militari,* translated from French into English by the command of Henry VII, London, 1489. Part of the contents is shown by Napoleon-Favé, *Études,* I, pp 17–45; II, pp. 3–71.

For the authentic title, *Livre des faicts d'armes,* 1410, see Jähns, *Geschichte,* pp. 347 ff. Jean de Meung, *L'art de chevalerie,* ed. U. Robert, Paris, 1897.

[5] Jähns, *Geschichte,* pp. 279–291; Berthelot (1891, 1900), *passim;* Sarton, *Introduction,* III, 2, pp. 1550 ff; A. R. Neumann in Pauly-Wissowa, *Real-encyclopaedie,* Supplement X, cols. 992–1020; White, *Medieval Technology,* pp. 32 ff. The text of the poliorcetics was published by M. Thévenot, *Veterum mathematicorum, Athenaei, Bitonis, Apollodori, Heronis, Philonis et aliorum opera,* Paris, 1693.

[6] *Thexaurus regis Franciae acquisitionis terre sancte de ultra mare necnon sanitatis corporis eius et vite ipsius prolongationis ac etiam cum custodia propter venenum* (Lat. 11015, Bibliothèque Nationale, Paris). It has been partly published by Berthelot (1900), pp. 416 ff; Uccelli, *Storia,* pp. 43 ff; Gille, *Les ingénieurs,* pp. 22 ff. A complete edition is planned by A. Rupert Hall, London.

tius. He knew it by heart and rerecorded it in his various works.

It is doubtful that Taccola knew much of other technical literature. Of Frontinus he knows the insipid *Stratagems* but not the book *De aquis urbis Romae*, discovered in his time. He seems unaware of the cistern constructions described by Palladius, *De agricultura* (I, 8 to 33) and of the oil press in Cato's *De re rustica* (XVIII to XXIII).[7] Of course the *De ingeneis* is silent about the surveying instruments of the Agrimensores or *Gromatici*, discovered by Cusanus only in the year Taccola completed his first treatise. Notably, he is silent also about Euclid and the ancient Archimedes; what little he knows of the science expounded by them seems to come through impoverished accounts in the so-called encyclopedias then current. In his numerous drawings of cranes, water scoops, and other lever devices he would have had occasion to show his knowledge of the peripatetic *Mechanical Problems*.[8] He shows none. He is interested only in arrangements of plain or double pivots, for vertical or horizontal swing, and in letting a winch drive levers to drive a pump.

Taccola's autographs never mention the name of Vitruvius, and they are silent about any and all of the technical teach-ings contained in his famous Ten Books, *De architectura*. At most some introductory and philosophical statements of the master have a kind of echo (I 36v, III 41v). The actual Vitruvian teachings, dealing with city planning, building materials, and the classic orders, were understood only by Taccola's successors in the field of theoretical writing. He is much closer to the medieval distillates from Heron and Vegetius than to any renaissance of Vitruvius. He does not show the slightest awareness of the Vitruvian house constructions, vaults, floors, finishes, paints, well surveying, aqueduct construction, and design of water clocks. Some splinters of this knowledge may be detected in the copybooks based on *De ingeneis*, for instance the first few folios of Add. 34113, but those books were written after Taccola's time. He does not even know the famous machines of the Tenth Book, with the result that his versions of machine elements, such as gear shafts, are inferior. Equally lost to him was the ancient abridgment written by Faventinus.

If Mariano nevertheless derives some knowledge from the Vitruvian work, as for example at III 43v, he receives it indirectly through the late ancient and early medieval *epitomae, claviculae,* and *schedulae artium,* minor extracts of former extracts. In using them and combining them with oral information and original thought, he contributes substantially to the gradual new beginning of rational teaching and accumulation of knowledge, which then gradually leads to the newer *trattati* of artist-theoreticians, *Büchlein* of practitioners, the later *Theaters of Ma-*

[7] Palladius, Rutilius Taurus Aemilianus, *De agricultura*, Venice, 1472; *Palladius on hosbondrie* (ca. 1420), ed. B. Lodge, London, 1873–1879. For other editions see Sarton, *Introduction*, I, p. 355. Cato, Marcus Porcius, Censorius, *De agri cultura*, in *Scriptores rei rusticae*, Venice, 1492. Other editions in Sarton, *Introduction*, I, p. 186.

[8] Aristotle, *Minor Works*, London and Cambridge, Mass., (Loeb ed.), 1936, pp. 327–411.

chines, and the modern textbooks of technology.[9]

It is clear that Taccola learned by looking and listening as well as by reading. He held, with Brunelleschi, that the architect must have read books and must also have seen the things, *multa legisse ac multa vidisse.* He used methods of experimental testing as well as observations of workmen. His claims of inventions have fair foundation, and his tests of 1427 led to further efforts of his followers. Perhaps his work produced tangible building structures, incorporating his teachings and improvements. However we must leave it to others to find such structures, for instance in the Val di Chiana, east of Siena, or in the coastal zones and harbors to the west.[10]

As noted in connection with IV 61r, the use of a powder mine for blowing up a fortress was long attributed to Taccola, although he lays no claim to it. The fact is that an ancient mining method was developed by other workers who incorporated the use of gunpowder in it and that the new method already was proposed for actual warfare in 1403.

[9] We refer to the Vitruvian *Epitoma* by Faventinus, third century A.D. (printed in large editions of the *De architectura,* see H. Krohn, *De Faventini Epitoma,* Berlin, 1896); the Carolingian *Mappae clavicula* (T. Phillipps, in *Archaeologia,* XXXII, 1847, pp 183–244); the *Diversarum artium scedula* by Theophilus, ca. 1100, ed. C. R. Dodwell, Cambridge, 1961; the *Feuerwerksbuch, Visierbüchlein* and similar outlines, produced ca. 1375–ca. 1550 (Sarton, *Introduction,* III, pp. 1551, 1580; Wm. Hassenstein, *Das Feuerwerksbuch von 1420 . . . Neudruck des Erstdruckes aus dem Jahr 1529,* Munich, 1943).
[10] G. B. del Corto, *Storia della Val di Chiana,* Arezzo, 1898; P. Marrara, *Storia istituzionale della maremma senese,* Siena, 1961.

Brunelleschi may have mentioned it to Taccola, who then produced the first illustrations of the new arrangement. An almost totally misleading impression is created when the first illustrator of a device is confused with its inventor, and the confusion is compounded when the illustrator's own statements, including his claims of 1427 (I 61r) to other inventions, are disregarded. Regrettably, this misconception or neglect of facts has dominated the accounts about Taccola written during the last century or two. One of the many scenes that he illustrated has been misinterpreted as showing a method introduced by him, while his truly significant work and explicit claims have been overlooked. Perhaps the strongest statement of this narrow and conventional view of Taccola, held by many scholars, is that of Reti (1964). He postulates that all pages of Taccola's work show devices that "are always in connection with the requirements, immediate or potential, of war. He did not have the divine curiosity of Leonardo, standing still before the marvelous patterns in the movements of air, water and flame; he has not Leonardo's unconfined interest in any shape or movement, in all creation, man-made or in nature." We share the admiration of Leonardo, but cannot agree with this indictment of Taccola.

The actual scope of Taccola's interest in the field of technology was almost as wide as the contemporary field itself. In this it resembled the wide scope of Leonardo's interest almost three generations later. Taccola illustrates more than the machines of attack and defense noted by his earliest rediscoverers. He also illustrates contemporary devices of civil and mechanical engineering. He deals actively, and in some respects creatively, with bridges and their foundations, with harbors, harbor equipment for the loading of freight, aqueducts both above and below ground, equipment for operating the wells at the end of an aqueduct, mill houses, and the machines and power plants associated with them. He also shows the elements of private dwellings and public buildings, including wall structures, chimneys, stairs, roofs, and possibly the layout of a facade.

The achievements of Taccola's engineering may not be as great as his interests were wide, but he contributed to progress. It would be foolish to expect from him, or from men greater than he, fundamentally new concepts. Vitruvius had defined well the possible achievements of an engineer's lifetime efforts and the aims of his own work (Book VII, Preface): to record the thoughts of former generations in such a way "that they, . . . being developed, . . . should gradually attain to the highest refinement," for this purpose "to draw from them and to go somewhat further" than they had gone, and in the process "to treat the different topics methodically." This is what Vitruvius actually did in his time, and this also is what Taccola did, beginning in many respects on a level lower than that of his great predecessor. His engineering work was that of an inspired teacher, one who was to develop the work of his predecessors and "go somewhat further."

Taccola did this, not only as a transmitter and originator of technical thought but also as a developer of methods of communication about technical thought. As to such methods, Taccola found little when he began, in spite of centuries during which the technology of Vegetius had been restated by many. There had been descriptions and drawings of machines, but there was substantially nothing in the earlier manuscripts—with the possible exception of Kyeser's attempts —to describe or suggest classes of machines, their common parts, their basic functions, or their general rules. Taccola tried to develop ways to suggest or describe or illustrate these matters. This led him to experiment with different forms of cross-reference between chapters, and this may also be among the forces that led him to develop different forms of illustration. Of course his forms and methods were soon surpassed. As noted by Reti (1964), only after Taccola are there machine drawings of "such technical excellence as only to be surpassed by Leonardo."

It was only long after Taccola, and for that matter long after Leonardo, that technical specifications began to resemble their present-day form, and we cannot be sure that they have reached a definitive form yet. Progress in all these respects had long been impeded, and still was impeded, by a number of factors, perhaps including the abstract and mathematicising tendencies inherited from some schools of ancient thought and the otherworldly tendencies of the Church Fathers. Also included among the retarding factors was the secretive attitude fostered by medieval guilds and early modern states, whose attitudes were accepted and even shared by the early engineers. Fortunately their attempt failed; natural curiosity is apt to overcome all attempted secrecy, and it was overcome in Taccola's case. He was fitted for clear speech, not for the "veiled" speech that he attempted.

As a writer, Taccola uses a medieval, notarial style. He shares with Vitruvius, or the Hussite Anonymus, or Leonardo, a distinct interest in conciseness. In this he differs widely from the bombastic style of Kyeser or the easy rhetoric of Francesco di Giorgio, Alberti, and their successors. Taccola always aims to specify the cardinal point; frequently he advises his reader to direct his mind to such a point. Eager to convey knowledge to his reader, he is neither eloquent nor systematic, but he is intense. When he makes a point, he makes it with force. We think he was a good teacher.

The Sienese were of the same opinion, and so were others in Italy. During his lifetime, Taccola taught hydraulics and mechanics—all that was then available— to a select few. He became so famous that after his death his manuscripts and their copies continued to be copied for as long as a hundred years. They were not then read as a historic account but as if they still were a living textbook. In this respect they were overestimated. They no longer had validity as live, technical information a hundred years after Taccola's death. The stream of modern development was far too rapid by then. Even the works of his younger contemporaries and early followers, like Leonbattista Alberti and Francesco di Giorgio, had begun to be surpassed by then. Leonardo's notes had been made,

dispersed, and probably forgotten in general. The day belonged to the generation of Vannoccio Biringuccio and Bartolommeo Neroni in Siena, San Gallo the Younger in Rome, Cardano and Tartaglia in northern Italy, and Agricola farther north. By then the process of transition to the modern era was very advanced, and the time of Galileo was approaching. Taccola's achievement, although slighter, was a pioneering work in the early phase of this transition.

MARIANO TACCOLA
AS ARTIST

7 Mariano's drawings in three autograph manuscripts document almost a lifetime of work. As long as neither Palat. 766 nor Lat. 28800 was considered, the phases of his artistic development could not be clearly defined. The record in Lat. 197 alone is too irregular to permit a clear pattern of his artistic growth. It contains a number of specific dates but a variety of drawing techniques in uncertain and ambiguous sequence. The newly identified relationships between the three autographs and their component parts resolve some of these questions. Each is a coherent unit, abundant in material that clarifies the styles of his middle and late years. Lat. 197 is interlaced with elements, both artistic and technical, leading toward Palat. 766 and Lat. 28800. Its contents and recorded dates partly overlap the evidence in the other two manuscripts.

Both the position and the drawing technique of the *capitula* (I lr–21v; 30, 31) indicate that this first part of Taccola's notebook is the earliest, ca. 1419. He draws confidently and presents essential elements clearly and concisely with firm outlines, usually without shading strokes. He sometimes uses a ruler and compass, yet his freehand drawing has similar clarity and order. A series of undulating pen strokes, at times ap-

plied densely, characterizes the surfaces and borders of bodies of water, for example in the harbor scene, I 13r (Pl. 1). Although pen strokes, of different quality are used in a variety of ways, the furry skins of a bull and the horse of the mounted gunner, I 21r (Pl. 3), are barely indicated; only in later drawings does Taccola emphasize these textures with detailed shading.

Landscape elements in the *capitula* are limited to the earth edges of river or harbors as indications of locations where engines are used or constructed. These sites are indicated by simple lines, flowing smoothly in broad curvilinear patterns, occasionally interrupted by a pointed projection. The rhythms are vigorous and bold, never intricate or subtle. The towers on shore and other buildings are equally direct, simple, and large in scale. An extended view of nature appears in the setting for a stratagem, showing a man in camp, I 21v (Pl. 4). It includes a windswept tree with foliage in pointed sprays, a type that recurs in Palat. 766 and Lat. 28800. The seated figure in this scene is drawn with apparent facility, the fluid contour line summarizing movements and textures with minimum means. The equestrianwarrior on the recto of the same folio (Pl. 3) is a much more static figure. These two drawings are indicative of the range of Taccola's technique during his formative years. Some are impulsive and imaginative, others rational and disciplined, and still others, such as the diver of I 31r (Pl. 6), fantastic or bizarre.

A second stylistic phase, which seems to have begun during or after the completion of the *capitula*, evolved naturally

out of the first. Perhaps the change occurred sometime before 1427, since that
year is recorded on two folios, I 42r, 61r,
coming relatively soon after the *capitula*.
Taccola worked on several books simul-
taneously, including a Book I comprising
folios 1 to 75 and a Book II, folios 76 to
96. Using the dated entries, cross-ref-
ences, and drawing styles as criteria,
we estimate that Taccola had begun
Book I and some folios of Book II ca.
1419–1425 and that he was well advanced
with them by 1427. His drawing technique during this second phase begins
with a fine boundary line, and this line
is articulated with delicate pen strokes
that are short, close-set, and slightly
hooked. He used these pen strokes sparingly and with control in application
and placement. In some examples attributed to this period, his modeling strokes
are longer, applied diagonally, and with
evident spontaneity, as for example
the drawing dated in 1427 at I 61r. All
drawings of this period have modeling
strokes on some parts more than others,
especially at earth edges, on voids, at
the edges of round forms or on planes
in perspective, and in places where
mechanical parts interconnect. This
technique, in conjunction with slight
crossing strokes, accentuates structural
details. The modeling technique seems
intended to aid in a clear showing of
parts arranged for relative movement.

Taccola's drawings do not reflect the
achievements of Brunelleschi in one-
point perspective. Throughout the work
Taccola consistently shows openings
of cylinders at independent vantage
points from above or from below, rec-
tangular or elliptical areas in indepen-

dent three-quarter views, and the thickness of surfaces by means of parallel
lines. He composes engines by shifting
from one or all of these views, while
projecting only the approximate position
or the connection of components. He
does not place mechanical or structural
parts in a rationally constructed space;
instead he shows elements separately.
He composes each engine and many
components from a different point of
view. Perhaps his technique could be
called multifocal.

During this second phase Taccola presents only one human figure, in the form
of Man as Microcosm, I 36v (Pl. 9), and
two horses used to drive engines; these
show little development over earlier
examples. There is equally little development in landscape; in more then twenty
drawings there is only the usual, cur-
sory schematization of earth edges tracing a lateral course or defining circular
pools of water. However, the form of
four landscapes (I 35v, 36r, 58v, 72r),
similar to the setting for a fulling mill,
I 40r (Pl. 10), goes somewhat beyond
these by showing a larger vista with
either rolling hills or low cliffs with
trees, plants on shore near bodies of
water, and buildings. Mariano develops
these vistas while he illustrates sources
and recipients of water for pumps and
mills and describes the interconnec-
tions.

A third phase is apparent in Books III
and IV (Pls. 21–108), which were probably begun in 1431. Here many more
elements are modeled with fine, close-
set pen strokes applied abundantly and
with deliberateness and delicacy. A
sensibility of touch and care for de-

scriptive details pervades each drawing. Each is complete and finished. These
refinements and others pertaining to
pictorial effects are probably due to the
experiments of intervening years. We
lack material, possibly recorded in lost
drafts or manuscripts, that would elucidate the steps leading to this phase.
Actually there are a few landscapes of
the third phase in Books I and II, shown
as settings for siphons, I 73v, II 94v
(Pls. 12, 18). As they are stylistically
similar to the group of landscapes in
Books III and IV, they may be dated to
the same years. Perhaps the siphon
drawings may even precede Books III
and IV by a year or two, since the later
books contain cross-citations to them.[11]

Books III and IV also bring new qualities of animation in these landscapes,
in human figures, and in animal life.
Plants or grasses and heronlike birds
appear near water. Natural formations
now include precipitous cliffs, mountains, and rocky plateaus, shown alone
or near water, and elsewhere combined
to form splendid panoramas with views
to distant mountain peaks. Taccola's
elaborate pen work has the effect of
making mountains appear hewn and
jagged, with crags and furrows running
vertically, and terraced bands rising like
steps. The windswept tree amounts to
a personal signature. He must have seen
quarries and split rocks and water
courses both gentle and precipitous. It
seems likely that we have here his response to what he saw when wander-
ing through the Tuscan countryside.

¹¹ All drawings cited in cross-references are
listed in note 8, p. 26 f; some drawings are
reproduced in Pls. III, 12, 15, 18.

At this stage in his development, Taccola includes much more animal life—
another sign of his interest in nature
study. Surely his drawings of the men
who manipulate his machines are studies from life. Their facial features and
body movements as they maneuver engines are individualized studies of Sienese workmen, as different from the men
in the *capitula* as from one another.
Taccola studied nature and studied her
carefully.

His continuing artistic growth is shown
in four drawings of figures and animals
on folios bound into Books I and II (Pls.
11, 16, 17, 19). On stylistic grounds these
and a number of others showing engines
may be dated shortly after 1433, and we
therefore call them Addenda. Their scale
is different from that in the preceding
books, and their modeling also shows
changes. For example, the ox working a
gin, II 96v (Pl. 19), is strongly foreshort-
ened, and its execution is relatively more
convincing than that of the working
animals in Books III and IV, which are,
with the exception of III 37r, shown in
profile. The anatomy is modeled vigorously, and the pen work is applied over
the entire surface. There is similar progress toward pictorial naturalism in human figures of the Addenda.

A relatively new method of modeling
in conjunction with larger scale is evident in the man starting a siphon, I 68r
(Pl. 11), and the soldier floating in water,
II 91r (Pl. 17):[12] pen strokes in straight

¹² This soldier and the rider on the facing
page, II 90v, recur in Taccola's *De machinis*,
Lat. 28800, 61r, 61v. Both are conceptually
known from earlier literature, for example
Frontinus, *Stratagemata*, III, 13; Guido da
Vigevano, *Thexaurus regis Franciae acquisi-*

parallel lines across one side of the body. Taccola had applied shading strokes of nearly similar character to the far side of the legs of animals in his vignettes of Books III and IV and used them sparingly in a few human figures of those books. Such strokes are more pronounced and regular in these two examples of the Addenda drawings, and they coincide with changes in proportions and anatomical form. The technique is not apparent in the rider shown at II 90v (Pl. 16), although the soldier on the facing folio 91r shares other stylistic features with the rider. As the new technique prevails in a large group of drawings, one folio of which shows the date 1438, it seems plausible to place the Addenda drawings ca. 1435.

The new technique is very characteristic of soldiers and engines in small scale, the Sketches inserted and annotated with small handwriting in the first two Books and in the Sequel. Sketches of engines, mainly military in function, may be seen on almost all pages of Books I and II; they are always arranged around primary drawings, often in copious array. There are hundreds of variant forms of ships, wagons, shields, bombards, and mountings. The shading technique in parallel lines and small modeling strokes are used, but the spacing of modeling strokes is wider and their application more casual. The execution seems rapid, sometimes so free and animated as to appear coarse. Dabs of ink in rapid succession generate visual rhythms and the effect of a frayed surface texture. Human figures, almost

all of them soldiers and riders, are shown in complex poses, and a sketchy rendering creates both figures and animals that are no longer stationary; instead they appear to move quickly. Descriptive detail as such is renounced; yet structural parts are clear and well located. The delicate, linear definition of form and the firm modeling that are characteristic of the third and fourth books of *De ingeneis* and the Addenda gives way progressively to greater abstraction in the Sketches, reaching a quasi-expressionism in figures, engines, and landscapes. These effects appear as sophisticated developments after Taccola's Books III and IV.

Who made these Sketches? Michelini Tocci (1962) postulates that Taccola could not rise above the primitive and wooden technique seen in many of his drawings. He suggests that either Francesco di Giorgio or some anonymous artist made the Sketches and that Taccola then added notations. We cannot agree. The refinements and high quality of Books III and IV suffice to disprove Michelini Tocci's theory, and the free pen work of the Sketches is a natural development from the earlier work. Inscriptions and notes appended to many of the Sketches are in Taccola's unmistakable handwriting; often they are applied in ways strongly suggestive of instantaneous writing at the very moment of drawing, not as additions to drawings made by another person. Pen work, ink, and rhythm of execution are often a clear unit. A number of the animals, although not all, show an accentuated shoulder joint that is a distinct trademark of Taccola, first used in the *capit-*

tionis terre sancte, Lat. 11–15, Bibliothèque Nationale, Paris.

ula. Engines throughout the *De ingeneis* and its inserts show the same features and oddities of wheel and shaft arrangements and are generally shown in three-quarter view. In view of our knowledge of Taccola's life and his announced secretiveness, it is most unlikely that he ever had an assistant for any length of time; and it is almost unthinkable that he had one for the length of time necessary for making the Sketches.

When were the Sketches made? As Berthelot (1891) noted, they must have been made after the larger contents of the same pages, the primary drawings. For a more specific answer, folio 82r (Pl. 14) provides evidence, as it has a large assortment of these Sketches on one page, and it contains two notes by Taccola stating that he showed the folio to two Sienese dignitaries on two successive days, the 8th and 9th of December 1438. One of the notes also cites certain types of bombards and military engines by type name, and these names recur on drawings of the engines shown on the same page: *Dominus Marianus Scizun de Sena die 8a mensis decembris vidit omnia ista in domo sue habitationis. Anno 1438 et ad 9 dicembre demo[stravi] domino Petri de Micheglis de Sena in designis bonbardam ad bossulam ad ciconiam ac ad vitem tunc dixit volebat immediate conferre un famulo Francisci Piccin[in]i etc.*[13] Evidently Taccola's free technique is well established at this time, five years after he completed Book IV. We think Taccola inserted the Sketches during years that began between 1433 and

1438. Perhaps he started making them ca. 1435. This expressive technique is then used by Taccola as late as 1449 in *De machinis,* Lat. 28800. Many engines of the Sketches—warfare apparatus, soldiers and their mounts—recur in large size in *De machinis.* They have the same textured effect, due to pen work applied swiftly and confidently. The technique is so similar in both works—the Sketches and *De machinis*—that we do not see how a more explicit chronology can be worked out. The year 1441 on folio 96r is not decisive in this question because it is merely an addition to drawings that existed on the page, and they have stylistic qualities typical of the Addenda. There are no further dates recorded in Lat. 197.

The exact starting point of the part that we have identified as a Sequel is not clear because there has been some dislocation of folios. This point was probably near original folio 99 (now unrecognizable), since numerals among the classified lists of pages, which refer mainly to Sketches in Books I and II, go to this original folio 99 and not beyond. The folios now remaining in this vicinity are 96 (original numeral unknown) and 97 (formerly 100). Although the Sequel and the Sketches are in different parts of the manuscript, and despite differences in figurative scale, these parts of the codex are closely related to one another and to *De machinis,* especially in the free style and sometimes in engineering content. The Sequel contains the Brunelleschi story (S 107v) and many drawings with texts of civil engineering and hydraulics, but military technology is resumed at folio 120. Draw-

---

[13] First transcribed by Berthelot (1891). *Scizun,* spoken *Sizzun,* seems to be Taccola's version of *Sozzin.*

ings in the Sequel and in *De machinis* are closely matched in style, and they share the subjects of mounted warriors and foot soldiers. Perhaps the Sequel drawings were conceived as part of, or perhaps preparatory drawings for, *De machinis.* As noted earlier, some sheets of paper in the Sequel have a watermark that is also found in *De machinis* paper. Regarding the problem of an exact relationship of these works, it is significant that Mariano repaginated a set of folios in the Sequel by placing a new set of numerals in the upper central margin rather than in the upper right corner;[14] he did the same to the *De machinis* paginations. Interestingly enough, some texts or lists in both works contain references to numerals of a former pagination series, not corrected to agree with the later series. In view of these and other complexities, a reasonable date for the Sequel may be ca. 1438–1449.

Stylistic changes developed after 1433 may be recognized in the Sequel. Landscape elements for the setting of machines, a typical one being that for a harbor chain, S 98r (Pl. 110), and others for hydraulic engines and construction devices, are executed with swift pen strokes as the Sketches of 1435–1438. In other Sequel drawings, soft and feathery strokes affect windswept trees and pine groves in ways that make them airy and decorative (S 98v, 99v, 112v, 113v), much less massive and sturdy than their earlier counterparts. These trees now dangle at

the edge of cliffs or are arranged on islands and peninsulas. In a total of thirty pages with landscapes, earth edges near water and the crevices of hillocks are now depicted as small undulations, varied and intricate in direction or recession, and modeling strokes articulate surfaces that have little physical substance (S 98r to 120r, mainly). In place of the deeply furrowed and vertical rock ledges of Books III and IV, many of the Sequel landscapes include mounds and hillocks swelling out of rolling hillsides and a horizon line that reaches to the upper edge of the page.

Human figures situated in landscapes of this kind are also affected by changes in technique and form. Among them, on ten pages, are construction workers and a fisherman. The proportions of these men and warriors, including foot soldiers and riders in S 123r and 136v (Pls. 111, 112), which are illustrative of the type, differ from the workmen of 1433. They are not as lean, and the anatomical articulation is loose. Their drapery, anatomy, and facial features are represented in greatly abbreviated form, summarized with a few pen strokes. Aside from a development in graphic technique, the Sequel shows technical change. The erroneous oblique position of a submerged buoy in Book I, 9r, 20r, has been corrected in S 98r (Pl. 110). Water controls, abundantly represented from folios 98 to 120, are shown in more conventional forms and with their conventional names, either Latin or Italian. The authenticity of the Sequel as a work of Taccola is shown by his handwriting on these pages and by the continuing stylistic and thematic elements of his

14 Folios 107r–118v of the Sequel are reproduced by Prager (1968), figs. 1–13. Regarding repagination numerals shown on some of these folios, see note 7, p. 26.

former work. The Brunelleschi story is an integral part of the Sequel.

Lat. 197 contains the nine sheets of a Separate Fascicule, SF 22–29, 133, bearing their former and consecutive paginations. There are no texts. These sheets show engine drawings of a coherent style, among them being a man operating a bucket wheel, SF 25v (Pl. 113). The modeling strokes are applied hastily and even carelessly; they are as loosely scribbled as those in some parts of the Sequel. The human operators in the Separate Fascicule resemble workmen of the Sequel in such details as the minimal indication of facial features, anatomical modeling, and proportions. However, their articulation is not as successful as it is in the Sequel. That the Separate Fascicule is nevertheless Taccola's work is fully manifest in the pen work, which is unmistakably related to that of the Sequel despite its declining form. In addition, there is identical pen work, although with more consistent expressiveness, and there are similar figurative types in the whole of *De machinis*. A number of the same engines recur there in almost the same sequence, accompanied by texts that mention workmen, which are not represented there but are shown in the Separate Fascicule. Perhaps Mariano wrote those texts on the basis of the Separate Fascicule drawings, which he had prepared but found unsatisfactory. The set of drawings may be an unused part of the *De machinis*.

Berthelot (1891) was right to call Lat. 197 both the draft and the extract of a treatise. Taccola worked from one book to another, placing a draft in one, a finished drawing in another, and vice versa. A set of sheets was moved from its original place. Folios 107 to 116 were first paginated as 228 to 237 in the upper right corner and were then repaginated by Taccola with numerals in the center of folios. As material accumulated and as new projects were conceived, he kept order in the record by means of cross-references and classified paginations. Parts of Lat. 197 are a treatise in the making, others amount to discards, and still others may be personal records made when some finished sets were slipped into a new context.

THE DRAWINGS OF
DE INGENEIS IN RELATION
TO SIENESE ART

Granting that Taccola's artistic development can be reconstructed only in approximate outline, between a few fixed dates, it is perhaps more definite than that now established for some of the well-known Sienese artists of this period.[15] Like other art in conservative Siena, his own work no doubt is explained largely by Trecento influences, particularly those of Simone Martini and the Lorenzetti. During his lifetime the Gothic style was modified in varying degrees by artists with whom he was personally acquainted—Jacopo della Quercia, the sculptor, Domenico dei Cori, master artisan of choir stalls, Do-

---

[15] Developments in Sienese painting are reviewed in Brandi, *Quattrocentisti*; G. H. Edgell, *History of Sienese Painting*, New York, 1932. Regarding sculpture, see Carli, *Scultura lignea senese*; P. Schubring, *Die Plastik Sienas im Quattrocento*, Berlin, 1907.

menico di Bartolo, painter—and by other contemporary artists of a new generation. Sienese art also received stimuli in Taccola's time through the work of famous visitors: Donatello, Lorenzo Ghiberti, and Gentile da Fabriano, among others. We can expect Taccola's drawings to disclose his reaction to such influences.

Taccola is an interesting and original Sienese artist. Even excluding machines, there are subjects among his book illustrations that are in part unexpected and unknown in Sienese art. There are human figures represented in situations seldom if ever portrayed in the major arts and nature studies among the vignettes near his texts that are distinctive to his art. His drawing technique shows an evolution from great refinement to impulsive freedom, traits of the true artist. Sienese art, including the art of Taccola, had inherited its manifold and partly conflicting traditions of stylization and realism in the presentation of nature and man. As Taccola achieved increasing power and freedom in his personal development and as he gained considerable fame in his professional life, he is likely to have influenced others, both artists and engineers.

MAN AS MICROCOSM

Among the earliest of Mariano's figure drawings is the image of man in a circle, I 36v (Pl. 9). It expresses the ancient and medieval concept of the Microcosm corresponding to the Macrocosm, and thus reflects the interrelated problems of measurement and proportions. Both were of interest to artists and theoreticians, and they continued to be dis-

cussed throughout the Renaissance.[16] Taccola, as one of these men and an architect, had good reason for including it in his book. He adds a builder's square, plumb line, and compass obviously in order to show that his version of the image is drawn by a teacher of the building arts. The drawing may have been intended as an emblem for a new chapter. His note on the drawing asserts a medieval point of view, and while it may not be his alone, yet it parallels the philosophical attitude he expresses in other drawings used as emblems, for instance the diver, I 31r (Pl. 6), and in texts about the architect and his work, III 41v. He writes, "He who knows all created me. I have all measure with me, of upper heavens, earth, and those below. He who understands himself understands much: the book of angels and of nature is in his mind [and that] of hidden things. Infra, etc."[17]

According to Vitruvius (III, ii, 2) and also according to Taccola's near-contemporary Cennino Cennini (Ch. 70), as well as other writers of the late Middle Ages, mathematical measurements predetermine the human figure and its proportions; they also reappear in other things of nature and art, the Macrocosm. These measurements are products of geometric forms such as circle, square, and triangle, commonly shown in Micro-

[16] R. Wittkower, *Architectural Principles in the Age of Humanism*, London, 1952, pp. 10 ff, Pls. 1–4.
[17] *Ille qui nichil ingnorat me creavit. Et omnem mensu(o)ram mecum habeo tam super celestium quam terrestium ac infernorum. Et qui se ipsum inteligit multa inteligit. Et librum angelicum et naturalem in mente eius habet asconditum:—Et infra etc.*

cosm figures of the Middle Ages and included here as well. It is uncertain whether Taccola knew Cennini's *Libro dell'arte*,[18] and it seems fairly certain that he did not know Vitruvius' books. Both authors had placed the navel of their Man as Microcosm at the center of the circle; Taccola places the genitals there. This modification—a substantial improvement in the interest of realism —is found in Villard de Honnecourt's pentagram scheme only if the head and other elements are disregarded;[19] yet it was explicitly stated, although not illustrated, by Lorenzo Ghiberti in the last portion of his *Commentarii*,[20] written about the time of Taccola's last years. Whether Taccola ever met with Ghiberti at any time in his life is unknown, and the relationship of their figurative concepts is unclear.

Literary sources on measurements designate either the human head or the face as a unit, without always making it clear what unit they use, and they vacillate between the use of multiples and common fractions.[21] The body length is given as 10 face lengths by Vitruvius or 9 faces in a Byzantine cannon and in Cennini's treatise. Ghiberti uses the head as module and writes of

$9\frac{1}{2}$ head lengths. Taccola differs from each of these. He divides the axis of the circle into 8 units, assigning one unit to the face. He uses 7 units for the height of the figure, beginning at the ankle joint. His divisions coincide in general, but not in every respect, with those parts of the body that are specified for a figure of 9 or more head lengths. The rules make no provision for containing the length of the feet when the figure is represented from an approximate frontal point of view, looking down on the feet, as it is here. The rules do refer to ankle joints and the arch of the foot, both of which are emphasized in the drawing. Taccola follows medieval-Byzantine convention in the position and outline of the feet; perhaps he even derives his image from a Byzantine manuscript.

Some of the curved and diagonal lines are used to fix locations of limbs and torso portions; for instance, two units of the module, equidistant from a point $1\frac{1}{2}$ modules below the top, establish the shoulders. The units are taken with a good measure of artistic license. The oblique lines on the right side of the figure seem to be by-products of the points of square, triangle, and axial subdivisions rather than guides to anatomical proportions. It is interesting that he treats the knee joints with the emphatic modeling that he also gives to the shoulder joints of his animals, I 30r, and throughout the vignettes of Books III and IV.

Features and proportions similar to those of Taccola's little man are seen in figures by other Sienese artists, particularly in work sometimes identified with

[18] *Cennino Cennini's Libro del' arte*, ed. D. V. Thompson, New Haven, Conn., 1932, p. 46.
[19] E. Panofsky, "The History of the Theory of Human Proportions as a Reflection of the History of Styles," *Meaning in the Visual Arts*, Garden City, New York, 1955, fig. 4.
[20] J. von Schlosser, *Lorenzo Ghibertis Denkwürdigkeiten ,I Commentarii.*, Berlin, 1915, p. 231; *Lorenzo Ghiberti, I Commentarii*, ed. O. Morisani, Naples, 1947, Book III (45), p. 214.
[21] Extracts of pertinent passages in the major sources are in Panofsky, "The History of the Theory of Human Proportions," pp. 55–107.

Gualtieri di Giovanni, a Pisan artist
active in Siena between 1409 and 1445,
who may be the same as the Master of
the Duomo Sacristy.[22] His figure of St.
Honofrius on an altarpiece (Pl. 114)
shows these proportions, in addition to
the same articulation of the knees that
Taccola uses here. Some of the figures of
Christ in the *Articles of the Creed* painted
by this master or by Benedetto di Bindo,
with whom he is closely related, are set
against a circular universe and His arms
are in positions as in Taccola's drawing.[23] One of these panels, a mere frag-
ment showing only the lower half of
Christ's body, contains feet that are
identical to Mariano's drawing in view
and delineation. The parallels are rather
numerous. The panels date ca. 1412;
they were executed, along with frescoes,
for the Cathedral, where Taccola had
made sculpture as early as 1408 and was
later to work again. The drawing of Man
as Microcosm is in the earliest part of
Taccola's Book I. Its severe outline and
rigidity contrasts with the graphic
fluidity of the man in camp, I 21v (Pl. 4),
the descriptive realism of the mounted

gunner, I 21r (Pl. 3), and a blend of realism and fantasy in the divers, I 30r, 31r.
The wooden quality of the image need
not indicate that it is earlier than the
capitula figures; it seems to be an exercise in methods of proportioning. In
any event each of these figures reflects
a particular facet of Taccola's artistic
impulses. At times he describes everything in detail; elsewhere he summa-
rizes a figure with a few rapid pen
strokes. The boldness and imaginative
qualities of his mind are also evident in
the drawing of caricatures and other
heads, which spin out of the initials of
his texts.

PORTRAITS OF SIGISMUND
Portraits of Sigismund were often made
by artists of his northern realms; some
were also made by Italians when the
king came to Italy.[24] Taccola made a
portrait for the frontispiece of his book,
III 27v (Pl. 22), and another, overlooked
by writers up to now, for the cover of
his manuscript (Pl. 20). In both drawings the king's facial features and ton-
sure are clearly those shown in portraits
of him by other artists, for instance the
well-known painting in Vienna and the
drawings usually attributed to Pisanello

[22] Some biographical data about Gualtieri di
Giovanni and questions of attribution are
discussed by Brandi, *Quattrocentisti*, pp. 33 ff,
181 f, in notes 15, 17. For Gualtieri's activity
in the Sacristy, and his relationship with
artists who painted the *Articles of the Creed*,
see Brandi in notes 8, 9, on pp. 175–177.
[23] Brandi published one panel of the series,
and two more are reproduced by G. Rowley
(*Ambrogio Lorenzetti*, Princeton, 1958, Pls. 134,
135), who attributes them to a follower of
Taddeo di Bartolo. The panels of the series
that are related to Taccola's drawing are unpublished. Photographs of them in the Frick
Art Reference Library show Richard Offner's
attribution to the Master of the Siena Duomo
Sacristy.

[24] L. Vayer (*Masolino és Roma*, Budapest, 1962,
pp. 135–158) reviews the literature regarding
portraits of Sigismund and a so-called Sigismund iconography. Additional portraits are
recognized in G. Szabó, "Emperor Sigismund
with St. Sigismund and St. Ladislaus: Notes
on a Fifteenth-century Austrian Drawing,"
Master Drawings, V, 1, 1967, pp. 24 f; G.
Scaglia, "An Allegorical Portrait of the Emperor Sigismund by Mariano Taccola," *Journal
of the Warburg and Courtauld Institutes*, XXXI,
1968, pp. 428 f

(1395–ca.1455).[25] Taccola achieves that degree of truth in lifelikeness that a portraitist usually aims to achieve. At the same time he achieves iconographic originality to a degree that is surprising for the early Quattrocento and is particularly remarkable in Siena.

In the frontispiece portrait God calls the king with the words, "Defend my sheep, whose custody I entrusted to you" (*Defende oves meas ex quibus te custodem elegi*). The king is shown as a warrior holding a sword and shield and listening to the word of God. He stands in an unconventional pose looking forward into the distance, perhaps inhabited by some enemy, his left foot on the tail of a seated, snarling heraldic lion. Since Sigismund was effective in putting an end to a war between Florence and Lucca-Siena, among his other Sienese activities, we may interpret this lion as the *marzocco*, symbol of the Republic of Florence. The beast resembles the marble *marzocco* carved by Donatello

in 1420,[26] conceived as a defender of liberty, not the traditional Florentine emblem of a walking lion, as may be seen on the pavement of Siena Cathedral. The portrait reflects a historic situation; through it Taccola makes an incisive, political statement and elevates his sponsor to heroic stature. He combines the figure of the king and a work of sculpture, which he may also have seen, and he formulates the character of the portrait on images he knew well.

Sigismund, tamer of the *marzocco*, is compositionally related to the figures included in the subject known as Famous Men of Antiquity. Taccola of course knew Taddeo di Bartolo's group of these heroes, painted in 1413–1414 in the Palazzo Pubblico in Siena,[27] and the figure of Joshua, another hero, engraved by Paolo di Martino in 1423 on the Cathedral pavement.[28] Perhaps Taccola intended to effect a symbolic parallel

[25] The painting in Vienna and the drawings in the Codex Vallardi, No. 2339, No. 2479, have been reproduced many times; all three are conveniently shown together in R. Brenzoni, *Pisanello*, Florence, 1952, Pls. 69, 71, 72. The attribution of the portrait in Vienna as the work of a Czech artist was made in journals not available to us (N. Rasmo," Il Pisanello e il ritratto dell' Imperatore Sigismundo a Vienna," *Cultura Atesina*, IX, 1955; R. Brenzoni, "Una rettifica attribuzione del ritratto di Sigismund," *Nuova rivista di varia umanità*, Verona, 1958.) Some doubt about Pisanello's authorship of one drawing in the Codex Vallardi is expressed by M. Fossi Todorow, *I disegni del Pisanello e della sua cerchia*, Florence, 1966. See also L. Magagnato, *Da Altichiero a Pisanello*, Venice, 1958; A. Schmidt, *Disegni del Pisanello e di maestri del suo tempo*. Venice, 1966.

[26] H. W. Janson, *The Sculpture of Donatello*, Princeton, 1957, Pl. 59-b and p. 41 f.
[27] For the iconography of Taddeo's figures, see N. Rubinstein, "Political Ideas in Sienese Art. The Frescoes by Ambrogio Lorenzetti and Taddeo di Bartolo in the Palazzo Pubblico," *Journal of the Warburg and Courtauld Institutes*, XXI, 1958, pp. 179–207. The first monograph on the artist is by S. Symeonides, *Taddeo di Bartolo* (Accademia senese degli Intronati. Monografie d'arte senese, VII), Siena, 1965. A medieval work on Famous Men, possibly owned by Taccola, may be found in the tract *De quatuordexe valentissimi huomini romani* transcribed in Lat. 7239, *De machinis*, fols. 125–129. The work is described in A. Marsand, *I manoscritti italiani della Regia Biblioteca parigina*, II, Paris, 1838, p. 3.
[28] The standard works about the pavement (R. Cust, *The Pavement Masters of Siena*, London, 1901; O. Turbanti, *Il Duomo di Siena*, Siena, n.d.) do not reproduce the Joshua panel.

with Taddeo's heroes of Republican Rome or the biblical Joshua. Inscriptions accompanying these Roman heroes in the Palazzo Pubblico extol their civic virtues and ask the Sienese to emulate them.[29] Sigismund's work for church and state, and the recognition it brought him at official and popular levels, must have determined Taccola's choice of figurative form. Furthermore, Taccola was in a position to know two prominent citizens, *Messer* Pietro de' Pecci and *Ser* Cristoforo di Andrea, who had directed Taddeo di Bartolo's fresco program and inscriptions. In 1413, *Ser* Cristoforo was sent as envoy to King Sigismund; *Messer* Pietro de' Pecci was one of the *priori* in 1426, *Capitano del popolo* in 1435, and doctor of law at the Studio.[30] Pecci, like Taccola, was named *comes palatinus* by Sigismund in 1433.[31]

If the figurative and symbolic prototype for the portrait of Sigismund is found in the cycle of Famous Men, its stylistic equivalent is the *Goliath* (Pl. 115) engraved on the pavement of the Duomo in 1423 by Domenico dei Cori (1363–1453).[32] Granting that the torsion of Goliath's body when hit by David's stone is inherent in the subject, nevertheless the figure shares with Taccola's drawing a number of features: a vigorous movement of the silhouette, drawn with a firm outline; a forceful thrust of the body, especially of the torso, with

some parts treated broadly and others, such as the armor, described in detail. Domenico dei Cori, master artist of the choir stalls in the chapel of the Palazzo Pubblico near Taddeo's frescoes, is one of the important Sienese artists with whose work other drawings by Taccola may be related, as we shall indicate presently.

Taccola's portrait of Sigismund is one of a small group of portraits of patrons on the frontispieces of books, as distinct from other forms that include portraits of authors or patrons receiving a manuscript from the author.[33] Not surprisingly, frontispiece portraits of patrons are of kings and princes. Sigismund's portrait belongs to a type—if it can be called that, since only one other example is known—showing the patron standing. Such a figure was produced in Padua in the context of dynastic biography. Taccola aspired to that art form when he stated in his dedication (IV 69r) that he wanted to serve Sigismund as biographer and illustrator. Taccola may have known the *De principibus Carrariensibus*, written ca. 1390 for the Carrara of Padua, which contains a portrait of Marsilio Maggiore da Carrara shown standing in profile.[34] The author, Pietro Paolo Vergerio, was educated in Padua,

---

[29] Rubinstein, "Political Ideas," p. 193 f.

[30] Biographic details about both men are culled from the Siena archives by Rubinstein, "Political Ideas," p. 190 notes 80, 81, p. 204 n. 164.

[31] *Regesta imperii*, IX, 2, p. 248

[32] Turbanti, *Il Duomo di Siena*, p. 71.

[33] C. Couderc, *Album de portraits des manuscrits de la Bibliothèque Nationale*, Paris, 1930. Also see H. Keller, "Die Entstehung der Bildnisse am Ende des Hochmittelalters," *Römisches Jahrbuch für Kunstgeschichte*, III, 1939, pp. 329 ff.

[34] It appears on c. 16v, not on the frontispiece, and is reproduced in G. Folena and G. L. Mellini, *Bibbia istoriata della fine del Trecento*, Venice, 1962, Pl. 21.

came to Siena in 1407–1408, and went on to work for Sigismund at the Council of Constance and in Hungary.[35] In addition to this source for frontispiece portraiture, possibly Taccola learned of such usage through Sienese artists who worked in northern Italy or perhaps through Veronese and Umbrian artists who were in Siena.[36] To be sure, the idea for it may be Taccola's spontaneous achievement, stimulated by his choice of a compositional model from among the cycle of Famous Men.

A second portrait of Sigismund is on the parchment cover (Pl. 20). Here the portrait is bust length and in profile. Little is known about portraits on manuscript covers, perhaps because they have worn away, as is almost true of this drawing.[37] However, there is a long tradition in Siena for narrative scenes and even papal portraits on the famous *Biccherna* account books.[38] Precedents for portraits of identifiable, living persons shown in profile-bust form, as is the case here, are extremely rare. This one ranks among the earliest of a type that became more common a few years later in the work of Pisanello.

The bust-length portrait of Sigismund is arranged in heraldic profile. The head is prominent; the shoulder is relatively narrow, simplified, and short, with perhaps a base line at the bottom. As a whole the bust is compact and rather tightly formed. These features make it resemble a sculptural bust or a medallion image without the customary framing device. It is about $4\frac{1}{2}$ inches high, executed in sepia ink in a tone slightly deeper than the drawings on folios. It is placed near the upper edge of the cover, somewhat off-center. The king's hair is finely waved and long, falling in regular pattern to the neck where it is trimmed straight. He wears a coronation crown, consisting of three diadems encircling a conical form and vertical strips joined at the top. Its total form has a general resemblance to the papal tiara and is clearly different from each of several crowns shown in scenes of the coronation ceremonies and other rites.[39]

---

[35] K. A. Kopp, "Petrus Paulus Vergerius der Ältere," *Historisches Jahrbuch*, XVIII, 1897, p. 541. Also see note 124, p. 127.

[36] Taddeo di Bartolo of Siena (d. 1422) worked for Francesco II da Carrara in Padua in 1389 and 1393. He introduced architectural forms of northern Italy and profile portrait busts in the framing borders of his frescoes in the Palazzo Pubblico chapel. Jacopo della Quercia was in Ferrara in 1408 and in Bologna in 1427. The style of the Umbrian painter, Gentile da Fabriano, was strongly affected by his experiences in northern Italy; he worked in Siena in 1425–1426. There is at least one Veronese artist, Cecchino di Francesco, who was active in Siena in 1432 at the time of Sigismund's sojourn. A record of his role as arbiter in the evaluation of Sassetta's *Madonna of the Snow* is known (J. Pope-Hennessy, *Sassetta*, London, 1939, pp. 25, 46 ff). Although little else is known about him, Longhi developed a theory (summarized by Brandi, *Quattrocentisti*, p. 181) regarding Veronese characteristics in the drawings of Giovanni di Bindino of Siena, dating before 1416.

[37] There is no mention of portraits on book covers in T. de Marinis, *La legatura artistica in Italia nei secoli XV e XVI*, Florence, 1960.

[38] A Lisini, *Le tavolette dipinte de Biccherna e di Gabella del R. Archivio di Stato in Siena*, Siena, 1901, Florence, 1904.

[39] The form of coronation crown shown in Domenico di Bartolo's engraving in the pavement of Siena Cathedral (Brandi, *Quattrocentisti*, Pl. 163) is not substantially different from that illustrated in the *Richental-Chronik* where Sigismund and Pope Giovanni XXIII are represented at the Council of Constance (*Storia di Milano*, VI, plate p. 176). Another

Evidently the crown is a product of
Taccola's imagination, created by re-
peating the diadem he had previously
drawn on the frontispiece portrait. In
that case, too, the crown worn by Sigismund is unique. The changed form of
the crown worn by Sigismund on the
cover portrait seems to allude to his
coronation in Rome, and it may be Taccola's later addition to his book.

The unexpected appearance of this form
of heraldic, profile bust for a recognizable, living person, at this early date
and in Siena, calls for a review of its
possible relationship with the work of
Pisanello. At some time ca. 1438, Pisanello produced portraits of North Italian noblemen and princes in the form
and proportions shown here.[40] Earlier
he had made at least one drawing, perhaps two, of Sigismund's portrait in
profile, when the king was in Italy in
1431–1433, probably on his return to
Germany via Ferrara and Mantua.[41] It
is possible, although undocumented,
that when Pisanello left Rome on 26
July 1432 en route to Verona, his journey
may have led him to Siena and that he

there made drawings of Sigismund
shortly after his arrival. In that event
it might be inferred that Taccola was influenced by Pisanello's form of portrait, although their respective drawings are unrelated either in technique,
conception, or ornamental detail. There
is no reason to doubt that the cover
portrait is Taccola's work, as the modeling strokes are those of his style throughout the manuscript and even the points
of the crown are identical to the diadem
of the frontispiece portrait.

It seems possible that Mariano and
Pisanello conceived portraiture of this
form and of Sigismund independently
at about the same time.[42] There is ir-
refutable evidence that Mariano's concept has antecedents in Siena, although
Padua may be the ultimate source. The
heraldic profile bust, based on Roman
coins and medals, came into use in the
last quarter of the fourteenth century in
Padua, Ferrara, and Verona for images of
Roman emperors and occasionally for
other persons of rank.[43] In the intellec-
tual climate stimulated by Petrarca and
through the influence of antiquarian
interests in Padua, the return to power
of the Carrara in 1390 was celebrated

form of crown occurs in Filarete's scene of
coronation (M. Lazzaroni and A. Muñoz,
Filarete, scultore e architetto del secolo XV,
1908, Figs. 61, 62), probably because the rite
called for at least three different crowns. This
is known through a description of the ceremony by Nicolo della Tuccia (*ibid.*, p. 71,
n. 4). For the ceremony in Milan, see *Storia
di Milano*, VI, pp. 267 f.
[40] See the bibliography for Pisanello in notes
24, 25, on pp. 169, 170, and some new observations on Pisanello's medals in R. Weiss,
*Pisanello's Medallion of the Emperor John VIII
Palaeologus*, The Trustees of the British Museum, 1966.
[41] These are the Louvre drawings, Codex Vallardi, Nos. 2339, 2479. Also see note 25, p. 170.

[42] It is interesting, although as yet unexplained, that Pisanello's medallion imagery
of the 1440s—portrait busts and equestrian
figures of north Italian noblemen—are incorporated in the drawings of the Paris copy
of *De machinis*.
[43] Portrait busts of Roman emperors, as shown
in manuscripts and frescoes, are best preserved in Veronese frescoes by Altichiero,
who was subsequently called to Padua to
work for the Carrara (G. L. Mellini, *Altichiero
e Jacopo Avanzi*, Milan, 1965, Pls. 1–13, 145,
297, *passim*). See also R. Pallucchini, *La pittura veneziana del Trecento*, Venice, 1965,
Fig. 694.

with commemorative medals bearing the image of Francesco II, clearly modeled on Roman prototypes.[44] A reflection of that antiquarian spirit is found in Siena in the borders of Taddeo di Bartolo's frescoes of Famous Men, painted after he had worked in Padua.[45] In addition, the Chronicle written and illustrated by Giovanni di Bindino before 1416—the same manuscript that records Mariano's account of the meeting at Radicofani—contains two drawings of his father, Bindino da Travale, in the form of profile busts bearing his name and framed in initial letters.[46] The personal acquaintance of Taccola and Giovanni di Bindino is a matter of record, and their respective literary-artistic works contain certain elements that distinguish them from the mainstream of Quattrocento Sienese art as we now know its form and character.

The fact that Sigismund was King of the Romans, considered the reigning successor to ancient Roman emperors, must have played its part in Taccola's selection of the heraldic profile bust so closely identified with ancient emperors. Perhaps Taccola made an intentional correlation between ancient and living men of virtue. Siena prided herself on her classical origin and her connections with ancient Rome.[47] Sigismund's presence in Italy and his far-reaching claims to a position as reformer of church and state had the effect of stirring artists to compose pictorial allegories. Some painters represented him as one of the biblical kings.[48] Indeed, Taccola may have spearheaded these innovations, possibly under the influence of learned men of his acquaintance, such as Pietro de' Pecci.

ST. DOROTHY AND THE CHRIST CHILD

Mariano's artistic originality prevails in his two drawings of saints in *De ingeneis*, subjects that were remote from his main interests and, in Siena, were least subject to change. His drawing of *St. Dorothy and the Christ Child*, IV 69r (Pl. 97), is consummately fine in execution and details, its beauty equal to the work of major Sienese artists. It is also a unique portrayal of the saint. Sienese painters, some of them Taccola's contemporaries, invariably follow Ambrogio Lorenzetti's precedent or the characteristically Italian iconography, by showing her with a lapful of flowers or a bouquet in hand, although they make minor compositional changes.[49] Taccola

[44] Weiss, *Pisanello's Medallion*, Pl. II, p. 11 f.
[45] Rubinstein, "Political Ideas," p. 192, notes that there are medallions in the borders framing Simone Martini's *Maestà* and in Ambrogio Lorenzetti's frescoes; however, they are not profile busts.
[46] P. Misciattelli, "Disegni inediti di una cronaca senese del secolo XV," *La Diana*, V, fasc. 1, 1930, Pls. 4, 5. As indicated in note 36, p. 172, a relationship with Veronese art was proposed by Longhi ("Fatti di Masolino e di Masaccio," *La Critica d'Arte*, XXV–XXVI, 1940, p. 186). Gentile da Fabriano, who came to Siena a decade later, may have helped to diffuse the concept, as he painted profile portrait busts (L. Grassi, *Tutta la pittura di Gentile da Fabriano*, Milan, 1953, p. 66).

[47] Details from contemporary sources in Rubinstein, "Political Ideas," p. 200 f.
[48] Scaglia, "An Allegorical Portrait," p. 429.
[49] Rowley, *Ambrogio Lorenzetti*, Pls. 36, 88. Italian paintings of St. Dorothy include Polyptych No. 200 in the Pinacoteca, Siena, attributed to Andrea di Bartolo, ca. 1420; a miniature illumination in the *Horae B. Mariae*, Ms A.R.7.3, Bibl. Estense, Modena; a triptych by Lippo Vanni, in the Lehman collection,

adopts the motif, but in a form different from these prototypes. In his drawing the saint is, unconventionally,
shown in complete profile; she holds a
bit of drapery with her hand in order
to carry the flowers in the folds of her
skirt. She is shown with the Christ
Child, clasping His right hand while
He holds a small pail or basket in the
other hand. Prior to this drawing, the
Christ Child had not been included in
Sienese or Italian images of St. Dorothy.[50] The motif of the Christ Child in
the iconography of St. Dorothy is known
in northern Europe, especially in Hungary, Bohemia, and Germany where the
cult of St. Dorothy was perhaps more
widespread than in Italy. Taccola may
have learned of northern symbolism at
the time of Sigismund's arrival and from
his staff; in any case he took it up with
originality.

A Bavarian woodcut print dating ca.
1410–1425, and a silver and enamel monstrance made in Basel, ca. 1430, depict
St. Dorothy holding the Christ Child
by the hand as they walk together,

while He holds a basket.[51] However,
nowhere in Northern art of this time or
later, nor in these two objects, is the
saint shown with a lapful of flowers.
This motif is purely Italian. In Northern
art her attribute is a basket, and the
Child does not always accompany her.[52]
Taccola seems to have synthesized elements of two traditions, and he does it
with style, giving his own interpretation to the story told in the *Legenda
Aurea*.[53] According to this legend the
saint was mocked at the scene of her
martyrdom by a scribe named Theophilus, who scornfully asked for roses
and apples from the garden of paradise. A child or angel then appeared to
her, bearing roses and apples in a basket,
and she asked him to bring it to Theophilus. She became a saint of gardeners
and florists, and people also prayed for
her aid in being spared the perils of
lightning, fire, thieves, and sudden
death. Perhaps Taccola, in letting her
speak for him and recommend him to
Sigismund, adopts her as a saint of
scribes, thereby alluding to his own
work and his offer to serve Sigismund

New York; *St. Dorothy* by Lorenzo Monaco,
formerly Larderel coll., Livorno (L. Bellosi,
"Da Spinello a Lorenzo Monaco," *Paragone*,
Year XVI, no. 187, 1965, Pl. 40).

[50] There is a possible exception, a painting
in Urbino showing the Madonna and Child
enthroned and St. Dorothy standing before
the Child, holding a basket of flowers which
He touches. Although the saint is traditionally identified as St. Rosa, the likelihood that
she is St. Dorothy is indicated by G. Kaftal,
*Iconography of the Saints in Central and South
Italian Schools of Painting*, Florence, 1965,
col. 368 and Fig. 414. See a sampling of images
of St. Dorothy in Kaftal, *Iconography of the
Saints in Tuscan Painting*, Florence, 1952, col.
329, and in L. Réau, *Iconographie de l'art
chrétien*, III: *Iconographie des saints*, I, Paris,
1958, pp. 403–405.

[51] *Europäische Kunst um 1400*, Vienna, Kunsthistorisches Museum, 1962, Pls. 25 (monstrance), 159 (Bavarian woodcut), and pp. 379,
285, respectively.

[52] Among others, see the following: A Stange,
Deutsche Malerei der Gotik, Munich-Berlin,
1961, XI, p. 149 f; X, p. 219; W. L. Schreiber,
*Handbuch der Holz- und Metall schnitte des
XV Jahrhunderts*, Leipzig, 1926, *passim*; H.
Fillitz, *Kunst aus Österreich um 1400*, Bad-
Vöslau, n.d., *passim*; A Matějček and J. Pešina,
Gothic Painting in Bohemia 1350–1450, Prague,
1956, *passim*.

[53] *The Golden Legend*, ed. Wm. Caxton, London-New York, 1900, VII, pp. 42–47; *Bibliotheca Sanctorum*, IV, Istituto Giovanni XXIII
della Pontificia Università Lateranense, 1964,
cols. 819–826.

as biographer. If he knew her cult was popular in the King's domains, as well as Siena, he may have wished to add to the bonds of friendship between the King and Siena.

Iconographically, the drawing is unusual, but its aesthetic qualities correspond with the prevailing late Gothic style as practiced in Siena by leading artists. The saint's drapery, richly modeled, falls in abundant cascade and breaks into intricate folds as it trails the ground. Her head in profile, the transparent veil, and the hair interlaced with ribbons recall the classic beauty of Ambrogio Lorenzetti's *St. Dorothy* and his painting of other women.[54] These elements were brought to great refinement in paintings by Taddeo di Bartolo, the leading Sienese artist of the early Quattrocento. Yet there are similar qualities in Gentile da Fabriano's St. Mary Magdalene in the Quaratesi polyptych of 1425[55] and even in the polychromed wood statue of the *Virgin Annunciate* in Montalcino (Pl. 116.) attributed to Domenico dei Cori.[56] Other Sienese artists, such as Benedetto di Bindo and Gualtieri di Giovanni who painted in the Cathedral and its sacristy ca. 1412, sometimes arranged their figures in profile,[57] but the concept is more prevalent in pictures by Taccola's young contemporaries. St. Dorothy may be compared with the lovely women in the Asciano altarpiece representing the *Birth of the Virgin* (1432–1436) by the Osservanza Master and in Sassetta's *Adoration of the Magi* (1432),[58] both of which show other elements derived from Gentile da Fabriano.

The drawing of St. Dorothy also discloses Taccola's awareness of a new fashion in dress that seems to have originated in the courts of northern Italy. It was probably then adopted by Sienese women but introduced into painting by Gentile da Fabriano, who had worked in Northern Italy. This fashion uses large fur cuffs on the upper arms and a roll at the neckline that is also perhaps made of fur. While the dating of relevant frescoes in Lombardy and Verona is problematic,[59] there is an exact date, 1425, for Gentile da Fabriano's Quaratesi altar painted in Florence, where angels are similarly robed.[60] Gentile was in Siena in 1425–1426, and perhaps earlier, when he painted the *Madonna of the Notaries*, now lost,[61] for the Ufficio dei Banchetti or office of notaries where Mariano read his reports for the Sapientia.

Taccola is among the first to display the new fashion in Sienese art, along with

[54] Rowley, *Ambrogio Lorenzetti*, Pl. 36.
[55] Grassi, *Gentile*, Pls. 76, 78.
[56] Carli, *Scultura lignea senese*, p. 31 f, proposes to date the statue group ca. 1405. Facial features and hair styles similar to this statue may be seen in the sculptures of Francesco da Valdambrino (*ibid.*, Pls. 93, 111).
[57] See the frescoes by these men and by Niccolo di Naldo in the Cappella degli Arliqui, the Sacristy, and the Cappella dei Liri in the Cathedral and others in the Carmine (Brandi, *Quattrocentisti*, Pls. 16a, 17a, 18a, 22–28, 35–39).

[58] Among other reproductions of these well-known pictures, see Brandi, *Quattrocentisti*, Pls. 62, 80, where the Asciano altarpiece is credited to Sano di Pietro, or Pope-Hennessy, *Sassetta*, Pl. IX, who gives the altarpiece to Sassetta.
[59] G. A. Dell'Acqua and F. Mazzini, *Affreschi lombardi del Quattrocento*, Milan, 1965, Pls. 50, 51, 60.
[60] Grassi, *Gentile*, Pl. 74 and p. 61 f.
[61] Grassi, *Gentile*, pp. 51 f, 66.

Sassetta who adopts it for angels in his *Madonna of the Snow,* 1430–1432.[62] It then obtained a conspicuous place in other paintings by Sassetta and by the Osservanza Master and Giovanni di Paolo in works of these years and slightly later.[63] We may add to these Domenico di Bartolo's *Madonna and Child and Angels,* signed and dated in 1433.[64] Thus an element of dress, introduced into Sienese art from North Italian courts, confirms Taccola's place among the new artistic currents of his city. Considering his official position in 1425, he may have known Gentile da Fabriano. Through St. Dorothy, Taccola recommends himself as miniaturist, and through her picture and those of the King, he displays his skill in its most refined form. It is also interesting that his innovation in the iconography of St. Dorothy underlies the composition of a small painting usually attributed to Francesco di Giorgio, in which the Christ Child appears with St. Dorothy

who holds a palm leaf.[65] If the attribution is correct, Taccola's influence on Francesco encompassed the design of devotional images as well as engines.

ST. GEORGE SLAYING THE DRAGON

The picture of *St. George Slaying the Dragon,* IV 76r (Pl. 108), with which Taccola closes his book, is still another example of his iconographic originality. St. George is the patron saint of metalsmiths and a warrior-saint who is identified with the church militant and with heroism. Taccola adopts the traditional figures in a landscape[66]—the saint and his horse, the princess, and the dragon near a cave—but he shows them in new form and relationship. Here the cave is an inconspicuous opening in the hillside, one that resembles settings for Taccola's engines; the cave was traditionally represented in a huge rock rising out of a plain and dominating the scene. In Taccola's picture the saint is dismounted, and the princess is chained to a column. We know of only one picture where St. George stands on the ground and is engaged in battle with

[62] Pope-Hennessy, *Sassetta,* Pl. III.

[63] See the dress of seated angels in the polyptych of San Domenico, Cortona (Pope-Hennessy, *Sassetta,* Pl. VIII), the dress of angels and women in the Asciano altarpiece of the Birth of the Virgin (*ibid.,* Pl. IX), which is now attributed to the Osservanza Master (E. Carli, *Sassetta e il Maestro dell' Osservanza,* Milan, 1957, pp. 96, 128 and plates) and now dated after 1436 by Carli. See also Brandi, *Quattrocentisti,* notes, pp. 197–201 for a history of the attributions, and Pls. 125, 126 for illustrations of the predella of the Fondi polyptych of 1436, where the women wear dresses of this type. Details of these and other pictures in J. Pope-Hennessy, *Sienese Quattrocento Painting,* Oxford and London, 1957, *passim.*

[64] Pope-Hennessy, *Sienese Quattrocento,* Pl. 38; Brandi, *Quattrocentisti,* Pl. 160.

[65] A. W. Weller, *Francesco di Giorgio, 1439–1501,* Chicago, 1943, Fig. 4; Pope-Hennessy, *Sienese Quattrocento,* Fig. 73. As both authors are unaware of Taccola's drawing, they believe the image is derived from German prototypes. The iconographic discrepancies are apparent in the comparison made by Weller. The Child in the German woodcut that he reproduces does not hold the hand of St. Dorothy, nor does He carry a basket; it is St. Dorothy who holds the basket. These are the motifs in Francesco's painting that seem certain to be derived from Taccola's work.

[66] O. F. Taube von der Issen, *Die Darstellung des heiligen Georg in der italienischen Kunst,* Halle, 1910.

the dragon. It is a small panel attributed
to Domenico di Bartolo, probably painted
ca. 1430.[67] It is otherwise entirely different from Taccola's version. Taccola gives
his own interpretation of the event, and
does not hesitate to use motifs of his
own that are at hand. The dragon appears earlier as the emblem to *Liber
secundus draconis* (Pl. 13) and is also used
as a vignette (Pl. 30). In the absence of
other scenes of St. George where he
confronts the dragon as shown here, it
is interesting to find that the scene has
some parallels with a miniature in a
French manuscript of *Lancelot du Lac*,
ca. 1380.[68] Like the hero Lancelot, St.
George wields his sword in a downward motion when the hand holds it
over head. But the saint's action of jab-
bing his shield in the dragon's jaw is a
motif conceived by Taccola.

TACCOLA'S LANDSCAPE
SETTINGS, CA. 1427–1433
The landscape settings of Books III and
IV and of a few drawings of Taccola's
first books (I 70v, 73v; II 94v; Pls. 12,
18) are innovations of his stylistic development beginning ca. 1427. Few Sienese
artists of Taccola's time, and none from
whom engineering work is known,
seem to be as responsive as he to the
manifold aspects of nature. Even fewer
artists illustrate the combined areas of
earth and water to the extent that he

does. In his exploration of technology
and physics, the natural background
has equality, if not a lead, in his technical pursuit. At least this is his position
in *De ingeneis* after a more conventional
beginning in the *capitula*. He devotes
special care not only to the peculiarities
of the setting and vista but also to its
poetic delights. There are large and
splendid panoramas for the site of an
aqueduct, III 32v–33r, a tide mill, III
34v–35r, and siphons over valleys and
mountains, II 73v, 94v, III 48v–49r, 56v–
57r. Among the hallmarks of his style
are the birds and plant life on shores,
windswept trees silhouetted against the
sky, and pine groves ranged along foot-
hill paths. With perceptive imagination
he illustrates the Tuscan landscape, its
valleys alternating with mountain peaks
reaching into space and visible in the
distance. Peaks rise up sharply and are
surmounted by fortress-castles, familiar
then as now in Tuscany.

Landscapes such as these readily recall the paintings of Simone Martini and
Ambrogio Lorenzetti in the Palazzo
Pubblico. Taccola seems to be one of
very few artists who views Tuscan landscape with their breadth of vision. His
landscape for an aqueduct (Pls. 32–33),
to cite one of many, compares with the
setting of Simone's fresco of *Guidoriccio
da Fogliano* (Pl. 117). In addition to topo-
graphical similarities, Taccola, like Simone, presents landscape elements against
an imaginary plane conceiving linear
rhythms gliding across it in crisp patterns. Fantasy and reality are closely
blended. Taccola's scenes are related
in other ways to Lorenzetti's vistas of
Tuscan hills, painted in the *Allegory*

[67] In the Edinburgh Gallery (R. Van Marle,
*The Development of the Italian Schools of Paint-
ing*, IX, The Hague, 1927, Fig. 342). This version is not included in the types known to
Taube von der Issen.
[68] A. Quazza, "Miniature lombarde intorno
al 1380," *Bollettino d'arte*, ser. v, Year L, 1965,
pp. 69–72 and Fig. 116 (Ms fr. 343, fo. 27v).

of Good Government.[69] Both artists represent nature as the place where man pursues his work. Perhaps the peasant who nudges his donkey, in Taccola's setting for a tide mill, III 34v, was adapted from Lorenzetti's version of similar figures walking through fields and city streets. Lorenzetti's small panel painting showing a harbor scene with a boat and a mountain close to shore is a unique precursor of Mariano's scenes with landscape and bodies of water.[70]

Taccola stylizes nature, as much as these and all contemporary Italian artists do, and like them he does so in forms of his own. The earth edges of rivers, pools, and other bodies of water are formed as linear undulations, boldly applied in regular rhythms. The surface texture of water is modeled with pen work that suggests its substance as well as its movement. He varies and magnifies earth edges of mountains and cliffs rising from plains or water in an obvious play of his imagination, so that each forms a new vista. His terraced plateaus are deeply furrowed in elevation; they are described by a severe and sinuous line that traces the upper plane, shown as tilted forward and parallel to the picture surface. The slopes of his mountains and cliffs are accented by

rock ledges in vertical bands, suggesting their height and varied levels as well as their hardness.

What are the sources of Taccola's conception of landscape? Men of his generation are masters of figurative art. Aside from Simone's and Lorenzetti's work, very few Sienese paintings with landscape settings antedate Taccola's drawings. A vista of mountain ranges surrounding a valley is placed in the background and subordinated to a row of large figures in the foreground in Taddeo di Bartolo's *Resurrection of the Virgin* (1408–1414) in the chapel of the Palazzo Pubblico.[71] Another, an intarsia panel for the choir stalls of the chapel made by Domenico dei Cori in 1415–1428, presents a somewhat more unified landscape as setting for the *Adoration of the Shepherds* (Pl. 118).[72] The rock setting near the hut and the earth rising gently to distant hills resemble Taccola's form of abrupt ascent, terraced planes, and furrowed slopes. There is no exact correspondence with Taccola's design of trees, but this may be due to differences in media and technique. Domenico's intarsia work is dependent upon a drawing, not necessarily his own, and it is the only scene among his panels that shows a landscape vista. It seems unlikely that Domenico dei Cori was Taccola's mentor in landscape imagery; the roles may be reversed. In any case, Taccola's conception of landscape probably came under the influence of a painter rather than that of a sculptor.

Among the few early paintings in

[69] A. Cairola and E. Carli, *Il Palazzo Pubblico di Siena*, Rome, 1963, Pls. XXXIV, XXXV, 42; Rowley, *Ambrogio Lorenzetti*, Pls. 222–232.
[70] Identified as a *Castle on the Shore of a Lake*, and its companion piece, *Citadel by the Sea* (Nos. 70, 71 in the Pinacoteca, Siena) in the most recent catalogue (*Guida della Pinacoteca di Siena*, ed. E. Carli, Milan, 1958, Pl. VI and p. 29). Rowley, *Ambrogio Lorenzetti*, Pl. 95, believes the lake scene is in the "tradition of Ambrogio Lorenzetti, in the mid-fifteenth century."

[71] Symeonides, *Taddeo di Bartolo*, Pl. LIX-b. Also see Pl. XIII-b.
[72] Carli, *Scultura lignea senese*, p. 29.

which landscape in its real sense plays an unusually large role are the predella panels of Lorenzo Monaco (1370–1425). Taccola's landscape for an aqueduct, III 32v–33r, is one of several that may be compared with Lorenzo Monaco's *Legend of St. Honofrius* (Pl. 119), dated ca. 1418–1422.[73] Both show terraced rock formations with vertical ridges, plateaus defined by sinuous outlines and tilted forward, and distant hills beyond valleys. In other panels by Lorenzo Monaco during these years—a *Nativity*, the *Legend of St. Nicholas of Bari*, and *St. Jerome*[74]—the elements of landscapes that recur in Taccola's drawings are applied in various arrangements: terraced rocks, outcroppings of rocklike mounds, stylized earth edges of a lake, and fortress-castles on distant hills. The extreme simplicity and clarity of Lorenzo's buildings correspond with these qualities in Taccola's aqueducts and other structures. Taccola may have known Lorenzo Monaco, a Camaldolite monk who worked mainly in Florence but who was Sienese by birth and whose Order had an important monastery near Siena.[75]

These landscapes by Lorenzo Monaco and Domenico dei Cori are conceptually related to Taccola's drawings and to the paintings of younger Sienese artists, all of which were made about the time of *De ingeneis*. Clearly Taccola cannot be the sole source of inspiration; yet he may have had influence as artist, not only as engineer. At the least he may have helped to evolve new combinations of landscape components or diffuse them or to be an intermediary who promoted an interest in landscape settings. His scenic motifs for engine drawings of 1427–1432 are datable to the years when a group of young Sienese artists gave the subject of landscape a more conspicuous place in traditional religious themes represented on panels.

The young, progressive artists of Taccola's time are Stefano di Giovanni, called Sassetta (1392–1450); Giovanni di Paolo (?1403–1482); the Master of the Osservanza and, formerly identified with him by some historians, Sano di Pietro (1406–1481). In their earliest pictures, the Master of the Osservanza and Giovanni di Paolo have a special affinity with Taccola. The rudimentary fortress-castles, as shown in Taccola's landscapes and in those of Lorenzo Monaco and Domenico dei Cori, are clearly different from the traditionally splendid and elaborate citadels represented in biblical scenes

[73] O. Siren, *Don Lorenzo Monaco* (Zur Kunstgeschichte des Auslandes, XXXXII), Strassburg, 1905, p. 98 f; G. Pudelko, "The Stylistic Development of Lorenzo Monaco," *Burlington Magazine*, 73, 1938, pp. 237–248; 74, 1939, pp. 76–81. For Lorenzo Monaco's drawings of the *Visitation* and the *Journey of the Magi* with similar landscapes, see Degenhart and Schmitt, *Corpus der italienischen Zeichnungen*, Part I, Vol. 4, Pl. 197.

[74] Pudelko published two pictures (in *Burlington Magazine*, 74, 1939, Pl. I), the predella panels in the Accademia, Florence. He sees Sienese characteristics in these panels and believes that Lorenzo's work influenced Sassetta and Giovanni di Paolo. The present whereabouts of the panel of St. Jerome is unknown.

[75] In Sta. Mustiola in the Sienese diocese (D. Johanne-Benedicto Mittarelli and D. Anselmo Costadoni, *Annales Camaldulenses ordinis Sancti Benedicti*, Venice, 1761, VI (1351–1430), pp. 240 f).

by artists of the Quattrocento who follow Trecento tradition in ways of their own. Such ornamental castles were represented by Taddeo di Bartolo, by Sienese artists of Trecento tradition, and also by Gentile da Fabriano.[76] Much simpler fortresses on mountain peaks, windtossed trees in pointed sprays, and animated bodies of water as components of panoramas were introduced into Sienese book illustration by Taccola and into Sienese painting at the same time by Giovanni di Paolo. Giovanni showed them first in the predella panel for the Pecci altar of 1426, representing the *Raising of Lazarus* (Pl. 120).[77] This lake scene, an isolated example of landscape in the predella as a whole, is treated more broadly than the rock cave in the foreground. Another artist, Sassetta, shows these fortresses on a strip of gently rolling hills, above the heads of figures, in predella panels for the *Madonna of the Snow* (1430–1432).[78] A characteristic of Giovanni di Paolo's landscapes, one that is fairly constant throughout his work and is distinctly different from Taccola's conception even when other elements may agree, is the arrangement of mountain peaks on an imaginary, flat base receding in the distance. Nevertheless, Giovanni di Paolo is one of the young Sienese artists with whom Taccola has close artistic connection.

In Giovanni's predella panels for the Fondi polyptych of 1436, the landscape includes windtossed trees and even a pine grove at the edge of a hill,[79] like those of Taccola at III 33r, IV 76r (Pls. 33, 108). In the *Flight into Egypt* of that polyptych (Pl. 121), he further introduces a large river flowing through his characteristic landscape, as well as other elements that are new in painting and are unprecedented in the subject. These components are unmistakably a part of Taccola's world: a millhouse in a river, a rustic landscape with workmen, birds in the sky, and fortress-castles on mountain peaks. Taccola shares with Giovanni di Paolo a strong reliance on a multifocal presentation, surprising in one who knew Brunelleschi.

The Master of the Osservanza also shows related motifs to Taccola's imagery in painting of these years. In one of four panels with landscapes in scenes of the *Temptations of St. Anthony*, variously dated as either 1432–1436 or 1429–1432, there is a lake with a ship and the shore dotted with seashells (Pl. 122).[80]

[76] See Gentile da Fabriano's *Adoration of the Magi* of 1423 in the Uffizi (Grassi, *Gentile*, Pls. 42–46, 65), Bartolo di Fredi's *Adoration of the Magi*, 1370–1380 (No. 104, *Guida della Pinacoteca di Siena*, Fig. 16), and Taddeo di Bartolo's *Funeral of the Virgin* (Symeonides, *Taddeo di Bartolo*, Pl. LVIII-b).

[77] C. Brandi, *Giovanni di Paolo*, Florence, 1947, p. 5; J. Pope-Hennessy, *Giovanni di Paolo*, London, 1937, p. 5 f.

[78] Outdoor settings for scenes of the founding and building of Sta. Maria Maggiore (Pope-Hennessy, *Sassetta*, Pls. V, VI; Carli, *Sassetta e il Maestro dell' Osservanza*, Pls. XII, XIII).

[79] In addition to the *Flight into Egypt* (Fig. 0), see the *Adoration of the Magi* in the Kroller-Müller collection (Brandi, *Quattrocentisti*, Pl. 125) and Giovanni's painting of St. Jerome on the Biccherna cover of 1436 (Brandi, *Quattrocentisti*, Pl. 127).

[80] The caption of our illustration shows Sassetta as the artist of the painting, following the explicit wish of the Robert Lehman Collection. We follow Longhi, A. Graziani, and Carli (*Sassetta e il Maestro dell'Osservanza*,

It is remarkable how closely it parallels the setting with seashells of Taccola's mill, III 34v–35r, or the siphon over a bridge, II 94v (Pl. 18). The motif of shells along a shore is sufficiently unconventional in art as to suggest a close connection between this artist and Taccola, and the sailing boat is another sign of Taccola's influence. He almost never shows a harbor or lake without adding a boat, as he does in the *capitula* (Pl. 1) and later in S 98r (Pl. 110). Two other panels by the Osservanza Master in the Jarves collection[81] correspond with Taccola's landscapes in such details as wind-tossed trees, valleys or paths amid rapidly rising hills, distant mountain peaks with fortress-castles, and simplified forms of architecture.

There is still another painting that seems clearly to reflect Mariano's influence. The landscape setting for the *Adoration of the Shepherds* [82] by an anonymous Sienese artist (Pl. 123) has a special affinity with Taccola's vistas in the combination of water, earth, and island. It recalls the landscapes of the siphon, II 94 (Pl. 18), and the aqueduct, III 32v–33r (Pls. 32–33). When these elements appear in combination with a heron standing prominently near the lake, there is little reason to doubt that Taccola's landscape imagery is the source of inspiration.

These landscape panels by the young Sienese artists, to which others of somewhat later date could be added, constitute with Taccola's drawings a group suggesting a close artistic, and perhaps an equally close personal, relationship. The relationship is not otherwise documented, but the evidence for it is not surprising in the light of Mariano's background as sculptor, friend of other artists, and holder of administrative positions that brought him some prominence in public. In landscape as in portraiture and images of saints, Taccola is evidently among the avant-garde in Siena, if he is not a major innovator. His landscapes are more varied than any by his colleagues, and they are a remarkably coherent series executed with skill and self-assurance. If exact proof of his influence is lacking, the fact that his later landscapes have continuing links with this circle of artists is cause for supposing that his drawings had influence and that it probably began soon after his appointment at the Sapienza.

THE WORKMEN IN
TACCOLA'S DRAWINGS

There are new forms and types of human figures in Taccola's third and fourth books. They appear in greater numbers than in Books I and II and in forms very new to the arts. In observing and representing working people, he may be guided by the examples of Ambrogio

---

pp. 89 f) who see stylistic differences separating the works of the anonymous master from those of Sano di Pietro and Sassetta who have been considered (by Brandi and Pope-Hennessy, respectively) as authors of these paintings. The date 1432–1436 was proposed by Pope-Hennessy (*Sassetta*, pp. 70–77); Carli suggests that it ought to be 1429–1432.
[81] Landscape settings for the *Flagellation of St. Anthony* and *St. Anthony Tempted by a Devil in Female Guise* (Brandi, *Quattrocentisti*, Pls. 98, 99).
[82] Previous attributions, which all seem unconvincing, are reviewed by F. R. Shapley, *Paintings from the Samuel H. Kress Collection. Italian Schools XIII–XVth Century*, London, 1966, pp. 147 f.

Lorenzetti's people in street scenes and countryside in the *Allegory of Good Government*[83] or by the familiar scenes of human life represented in other Sienese paintings describing legends of local saints, the building of Noah's ark, and other popular themes. Each of Taccola's figures is sufficiently uncommon to make it certain that he directly studied Sienese people at work. Men turning handles of horizontal shafts are shown from the back with hands uplifted (Pl. 25) and from the front with hands down (Pl. 53). The body motions, governed by the machine structure and governing the machine operation, are illustrated. Operators of vertical capstans, such as those of III 50r, 51r (Pls. 62, 64), come forward with hunched shoulders as they push the *timone*. The miller who has brought the crank of the millstone to its dead point near him, IV 65r (Pl. 89), is beginning to push it sidewise and back; his hands, firmly gripping the crankrod, are still in dead-point position, but his body is swinging toward the machine. The surveyor, IV 58r (Pl. 75), peers most attentively. The man who pulls down the force arm of a heavily loaded lever, IV 62r (Pl. 83), seems to do so with genuine effort and is unconventionally shown from the back.

While the figures are compositionally his own, they have formal elements indicative of their place in historical time. Their small stature and slender proportions are similar to those of men represented by Domenico dei Cori, Paolo di Martino, Sassetta, and others.

The intricate folds of long robes and their rich modeling accord with late Gothic style. Whether the dress is long or short, it has peculiar and complex creases at the waist, a motif that also occurs in earlier, wood sculptures by Domenico dei Cori and Francesco da Valdambrino.[84] Similarly, the purely arbitrary fold at the neckline of workmen's tunics is a common device in sculptures by these same artists. Quite uncommon, however, is the handling of the tunics of the man pulling a lever and of St. George (Pls. 83, 108) as agitated, rippled surfaces appropriate to the vigorous movement of their wearers. It is a Lorenzettian motif, seen in the dancing women of the *Allegory of Good Government,* and rarely used thereafter, except by Bartolo di Fredi in a figure fleeing and by Sassetta in a mason working a sieve.[85]

Other engineers or their draftsmen, such as the artists working for Kyeser,[86] had shown workmen as well as landscapes to indicate the dimensions and functions of their machines. However, we know of neither engineering books nor paintings before Taccola's that have his variety of individualization and refinement in portraiture, including the portraits of workmen. *De ingeneis* is part of a continuing Sienese tradition

[83] Cairola and Carli, *Il Palazzo Pubblico*, Pls. XXXI; Rowley, *Ambrogio Lorenzetti*, Pls. 206–208, 222–234.

[84] Carli, *Scultura lignea senese*, Pls. 66, *passim*.
[85] Bartolo di Fredi's fresco, *Earthquake in the House of Job* in the Collegiata at San Gimignano, ca. 1356–1367 (E. Carli, *Sienese Painting*, Greenwich, Conn., 1956, Pl. 85); Sassetta's panel representing the *Building of Sta. Maria Maggiore* (Pope-Hennessy, *Sassetta*, Pl. V; Carli, *Sassetta*, Pl. XIII).
[86] Cod. phil. 63, Göttingen, Bibl. Univ. See the publication by Berthelot (1900); *Bellifortis, passim*.

of narrative realism, which has further development in the portraits of professional and common men in the Pellegrinaio of the Spedale della Scala by Domenico di Bartolo.[87] Was Domenico influenced by Taccola? The two men, as mentioned in the biographic account, had some contact.

NATURE STUDIES.
"LOMBARD" MOTIFS AMONG
THE VIGNETTES

Taccola presents a great variety of zoomorphic creatures and a few human figures as decorative vignettes for his texts in Books III and IV. The arrangement itself is unusual; it is not used systematically in Taccola's other manuscripts, although occasionally there are caricatures and other heads near the texts of *De machinis*. The vignettes serve the author in matters of cross-citation and give charming ornaments to the pages. As book illustrations, their fine execution and the variety of birds and animals corroborate Taccola's statement in the dedication to the effect that he would serve Sigismund as illuminator of manuscripts. These creatures of nature are more numerous and varied than the domestic and fantastic animals of Taccola's earlier and later books. Aside from birds and beasts—domestic, wild, or fabulous—he includes a nude putto riding a toy stick, an equestrian figure, and the symbol of Siena, the she-wolf and twins (Pl. 34). His drawing of fishes in III 54v is accompanied by a proverb; the "giraffe" at IV 61v is identified *girafa*. The camel at IV 70v is dated 1432.

These drawings are remarkable achievements for the high degree of accuracy and variety in the rendering of animals and birds. With modeling strokes Mariano succeeds in capturing the effects of hair and feather and through outline the lineaments and proportions of the species. Direct studies of nature—particularly the birds, and the domestic and wild animals of the region—at least partly supersede his earlier, formalized images of the horse, bull, and fish in the *capitula*. Taccola, like other artists, usually shows walking and running beings in profile; now he represents some animals and birds in the vignettes, and also the horse operating an engine and birds near water, coming forward or poised to do so. He so shows many of his workmen operating engines, as well as a nude putto and an equestrian among the vignettes (Pls. 36, 38), and—perhaps somewhat later—an ox operating the ox gin (Pl. 19). In these examples Taccola undertakes foreshortening boldly and achieves it repeatedly. For his animal drawings he continues to use the earlier, strongly modeled shoulder joints, which is a mark of his style.

Nature studies of birds and animals, as distinguished from the domestic animals shown as operators of engines, were popular at this time at the courts of northern Italy, especially in Lombardy with its artistic centers in Milan and Verona.[88] Taccola wrote his books not

---

[87] Brandi, *Quattrocentisti*, Pls. 167–175.

[88] The basic study of the subject is P. Toesca, *La pittura e la miniatura nella Lombardia*, Milan, 1912, on which even the most recent evaluations are based: *Arte Lombarda dai Visconti agli Sforza*, Palazzo Reale, Milan, 1958; Magagnato, *Da Altichiero a Pisanello*, 1958; M. Salmi, "La pittura e la miniatura gotica," *Storia di Milano*, VI, pp. 785–855. Also see van Marle, *Development*, VII, 1926, Figs. 33–55,

long after the first of such studies ca. 1390 by Giovannino dei Grassi.[89] Other northern artists of this genre, namely Michelino da Besozzo, Stefano da Verona, and Pisanello were actively engaged as artists in the north while the *De ingeneis* was in progress.[90] The portable works of these artists, like those of Taccola, are known only in princely manuscripts or personal notebooks. This makes it difficult to assess the exact avenues by which Taccola—alone in Siena—came to respond to similar motifs, or to decide whether he knew the work of these men. Sketchbooks traveled with the artist. None of these men, and none of their books, is known to have come to Siena, and no other Sienese artist is known who shared this interest. Taccola may have been stimulated in the subject of animal portraiture by the examples of Gentile da Fabriano, who borrowed from Lombard or other northern art such motifs as hunting dogs and wild animals and introduced them into the landscapes of religious scenes. However, neither Gentile nor his Sienese followers show creatures of the types used here by Taccola. There is no proof that Pisanello accompanied Gentile to Siena, and although the only painting signed by Michelino da Besozzo is in Siena, its history and the artist's travels are quite unknown.[91] Perhaps the painter Cecchino da Verona, who resided in Siena in 1432, was influential in diffusing these subjects.[92]

Domenico dei Cori, with whom Taccola was associated, had a book on birds, lent by Lorenzo Ghiberti to the goldsmith Ghoro di Ser Neroccio sometime before 1425.[93] There was also abundant visual and descriptive material in the scientific literature and encyclopedias of natural history, the *Taccuinum sanitatis,* and other illustrated treatises, both western and Byzantine.[94] As scholar and writer at the Sapientia, Mariano had access to manuscripts, to scholars and scribes, and to illuminators.

70–76. Marginal drawings of animals and plant life near texts are used by Marco di Bartolommeo Rustichi in the Codex Rustichi (Degenhart and Schmitt, *Corpus der italienischen Zeichnungen,* IV, p. 376, figs. 505a-d).

[89] Painter, master of the workshop, and sculptor of the Cathedral of Milan, ca. 1390, Giovannino de' Grassi's activity and the identification of his drawings were first made known by Toesca. The art of Michelino da Besozzo shows his influence; Stefano da Verona and Pisanello are not entirely unaffected by his work.

[90] In addition to the literature cited above for these North Italian artists, see O. Pächt, "Early Italian Nature Studies and the Calendar Landscape," *Journal of the Warburg and Courtauld Institutes,* XIII, 1950, pp. 13–47.

[91] Pope-Hennessy (*Sassetta,* p. 95 n. 55) thinks Michelino actually visited Siena. The *Marriage of St. Catherine,* the only work signed by Michelino, was already in Siena in 1816, and certain frescoes in Siena seem to be explicable as the result of contact with some Lombard artist analogous to Michelino. Also see Magagnato, *Da Altichiero a Pisanello,* pp. 30–32.

[92] See note 36, p. 172. With regard to his later appearance in Verona and some works attributed to him in recent years, see Magagnato, *Da Altichiero,* p. 63.

[93] Milanesi, *Documenti,* II, p. 120.

[94] There are many or full illustrations in the following. E. Berti Toesca, *Il Tacuinum sanitatis della Bibl. Nazionale di Parigi,* Bergamo, n.d.; L. Serra, *Theatrum sanitatis, Codice 4182 della R. Biblioteca Casanatense,* Rome, 1940; *Tacuinum sanitatis in medicina,* Graz, 1965; *Storia di Milano,* VI, *passim.* Taccola probably knew one of these manuscripts, for his sketch of a carcass strung up on a beam (Lat. 197, 74v) strongly resembles the manuscript versions.

Some animals and birds are common to the decorative borders of manuscript illuminations produced in many schools in Italy and throughout Europe, and medieval bestiaries have a long history. Greater details than this we do not know at this time, since no study has been made of Sienese manuscripts except for the later work of Liberale da Verona and Girolamo da Cremona.[95] It is clear, however, that both observation of nature itself and literary or graphic sources provided Taccola with material for his nature studies.

Taccola's nature studies reveal his unusual independence of vision, especially when we find a possible source for his subject among the drawings of Giovannino dei Grassi or other Lombard artists. His vigorous modeling of textures and his treatment of the shoulder joints are distinctly his own, as are some modifications in anatomical form. The "monkey" holding a mirror may owe something to Giovannino's drawing of this subject in the Bergamo codex (Pl. 124), although it is not clear whether Taccola ever saw his drawings; in any event Taccola makes a significant change in transforming a monkey's head into a human caricature. That he did so instinctively is evident in sketches of faces or caricatures created out of pen flourishes in the initials of some texts. Taccola's vigorous pen work usually succeeds in creating powerful forms from sketchy archetypes in the work of Giovannino dei Grassi. Suffice it to compare the animals on this sheet from his Bergamo codex with Taccola's bear cubs, the

hedgehog, and the walking monkey (Pls. 51, 78, 80, 90).

Some animals, such as the unicorn, wild boar, leopard, hunting dog, and running stag (Pls. 40, 44, 53, 74, 76), are adaptations from images in Lombard or other sketchbooks and manuscripts.[96] Model books, such as the one owned by Ghiberti, and nature itself are sources for his birds, lamb, squirrel, the donkey nudged by a peasant, and the rabbit strung to a tree. Perhaps he drew from life the snake swallowing a toad, the cat with a kitten in his mouth, and the fox with a dead duck (Pls. 73, 86, 88). Yet the latter motif has a counterpart in the sketch of a bear holding a dead duck, attributed to Stefano da Verona.[97] In drawing other animals Taccola works with facts and fantasy. Not all are successful, but most of them are original. His "giraffe" (Pl. 82) is surely dependent upon accounts he had heard or read that described how its long neck permits it to reach the fruit of a date palm, which he takes care to show.[98] Poor information or lack of information may explain why its head and neck resemble his crawling dragon (Pl. 56) and the body is that of a donkey. His drawing of a camel (Pl. 100) differs from the traditional forms as seen in reliefs by

[96] *Arte Lombarda dai Visconti agli Sforza*, *passim*; Toesca, *La pittura*, *passim*; van Marle, *Development*, VII, Figs. 44, 46, 62.
[97] Codex Vallardi, fo. 195 (No. 2398) illustrated in Magagnato, *Da Altichiero*, Pl. XXXXIX. The sketch occurs on a folio containing drawings for Stefano's *Adoration of the Magi*, signed and dated 1435.
[98] Faulty representation of exotic animals such as this one may have induced Ciriaco d'Ancona to send a drawing of the giraffe in his letter to Taccola (*supra*, p. 16).

[95] A record of Sienese and foreign illuminators is given by Milanesi, *Documenti*, II, pp. 380 ff.

Jacopo della Quercia and in Bartolo di
Fredi's frescoes in San Gimignano in
that hair is added along the underside
of the neck. The elephant eating apples
(Pl. 92) recalls similar versions in Lombard sketchbooks and manuscripts.[99]
However, Mariano's elephant has a
curled trunk and the legs of a donkey,
a bizarre combination probably due to
Taccola's lively imagination. Similarly,
the fabulous griffin and the dragon or
wivern are plentiful in medieval art.
Taccola creates his own versions even
of these familiar emblems. He copies
his own work by using the emblem of
Book II (Pl. 13) for the dragon or wivern
in the vignette and in the scene of St.
George (Pl. 108).

Taccola's predilection for creating variants of even the most familiar motifs
is noticeable in the symbol of Siena
(Pl. 34). By changing the position of the
twins in this famous group, his emblem differs from each of many Sienese
versions extant in sculpture, frescoes,
intarsia, reliefs, mosaics, and metal-
work.[100] His ability and readiness to
modify an archetype may also explain
why the nude putto riding a toy stick
(Pl. 36) differs substantially from models
that occur in a Lombard sketchbook and
in the treatise on mechanics by Konrad
Kyeser.[101] Taccola shows him as a plump

child standing in a position to come
forward, while the Lombard artist and
Kyeser had represented putti only in the
simpler profile view running with this
toy. Taccola depicts less apparent agita-
tion but suggests more real substance
in the nature of potential action.

TACCOLA'S LATE STYLE.
FURTHER RELATIONSHIPS
WITH SIENESE ART

Taccola's later drawings in Lat. 197 continue to show connections with the art
of the same circle of Sienese artists.
That the parallels are fewer is due to
the nature of Taccola's subjects in the
Sketches, the Sequel, and the Separate
Fascicule. His figures are almost all soldiers; Sienese painting has little place
for these in religious subjects. His landscape panoramas are individualized in
accordance with the needs of hydraulic
engines, while those of Sienese painting
are traditionally limited to a few explicit subjects. Even so, certain elements
of Taccola's late landscapes and figures
in the Sequel recur in some Sienese
paintings, the dates of which are fairly
well established. These paintings confirm
the dating proposed for the later parts
of Lat. 197, based on stylistic grounds
and on internal evidence. As mentioned
earlier, the changes in Taccola's technique of the late 1430s coincide with
new effects. His late landscapes—twenty
of them—are conceived with delicately
turned earth edges, rolling hills and
mounds that leave little strips of sky,

---

[99] Toesca, *La pittura*, Fig. 261; van Marle, *Development*, VII, Fig. 72.

[100] In addition to the well-known representation on a column, there are others in various materials in the chapel and antechapel
of the Palazzo Pubblico and in the pavement
of the Cathedral.

[101] Folios from the Lombard sketchbook in the
Morgan Library are now reproduced, and
called Neapolitan ca. 1370, in Degenhart and
Schmitt, *Corpus der italienischen Zeichnungen,*

Part I, Vol. 3, Pl. 132. Also see J. von Schlosser,
L'arte di corte nel secolo decimoquarto (Raccolta pisana di saggi e studi, 13), 1965, Pl. 42.
For Kyeser's drawing of the putto, see *Belli-
furlis*, fo. 94r.

and textural surfaces of somewhat thin
and transparent character. The harbor
chain at S 98r (Pl. 110) is characteristic
of this concept. Comparable elements
and effects are seen in Giovanni di
Paolo's *Christ in the Garden* of 1436–1440
(Pl. 125). In addition to a high horizon,
there is a pictorial fusion of plains, swol-
len hills, and mountain ranges finer
than Giovanni had previously painted,
and rivers that move down hills or
through a valley. These topographic
features are perhaps more typical of a
group of pictures by Sano di Pietro and
the Osservanza Master. Their landscapes
in three panels representing *St. Jerome
in Prayer*, two of which were painted in
1436 and in 1444,[102] show the horizon
elements that are characteristic of Taccola's landscapes, including tree forms
and fortress-castles.

Turning to the foot soldiers and riders,
S 123r, 136v (Pls. 111, 112), among others
of this kind in the Sequel, their arrangement in well-spaced rows suggests that
Taccola drew them from his own or
another manuscript. That he used his
own work is evident in the gunner,
S 126v, who is adapted from an earlier
version I 21r (Pl. 3). As the series of
riders illustrate combat positions and
use weapons maneuverable in those
positions, his source may have been the
Stratagemata of Frontinus or, most likely,
an illustrated treatise on horseman-
ship, such as the *Fior di Battaglia* written
in the early fifteenth century by a Venetian teacher, Maestro Fiore dei Liberi da
Premariacco.[103] The combat positions

shown in the Venetian treatise are similarly represented by Taccola; but in
addition to strong stylistic differences,
the riders and their mounts are shown
wearing armor, while this is not the
case in the Venetian treatise. However,
the Venetian riders have crowns placed
on their heads or near those of horses,
a device extensively used in *De machinis*,
where many of these riders recur, and
once in *De ingeneis*, I 14r. Mariano's interest in armor and weapons is displayed
in the elaborate finials on the helmets
of some foot soldiers and riders (Pls.
111, 112). He probably saw *condottieri*
wearing them. They are part of the rich
pageantry of effects in Uccello's *Battle
of San Romano*, which celebrates a major
episode in the Florentine-Sienese war
mentioned by Taccola, IV 73v. Taccola's
foot soldiers have still another parallel
in Giovanni di Paolo's *Christ in the Garden* (Pl. 125), showing a troop of soldiers
crossing the countryside. They compare
with the soldiers in S 136v (Pl. 112)
and with a lantern bearer among the
Sketches, I 30r, in the way each soldier
steps forward.

The diffusion of these motifs of landscape and of soldiers in contemporary
painting may perhaps be traced to Taccola's influence. His landscape vistas
occur in much greater variety than those
of Giovanni di Paolo and Sano di Pietro,
and he had begun to develop them in
work predating their paintings. It is

---

[102] Brandi, *Quattrocentisti*, p. 79, Pls. 91, 92,
93, 108.

[103] The manuscript (Morgan Library, M. 383)
contains 107 small pen-and-ink drawings by

two Venetian or Veronese artists, followers
of Altichiero (*Italian Manuscripts in the Pier-
pont Morgan Library*, ed. M. Harrsen and G. K.
Boyce, New York, 1953, Pl. 35, p. 27). Other
reproductions from this manuscript in Schlosser, *L'arte di corte*, Pls. 36, 37.

certain that he is independent of these artists, and it seems possible, or even plausible, that the paintings from this circle of Sienese artists reflect Taccola's influence. We see him as an active developer of many motifs, notably those of nature herself, and as a significant contributor to the continued evolution of Sienese art.

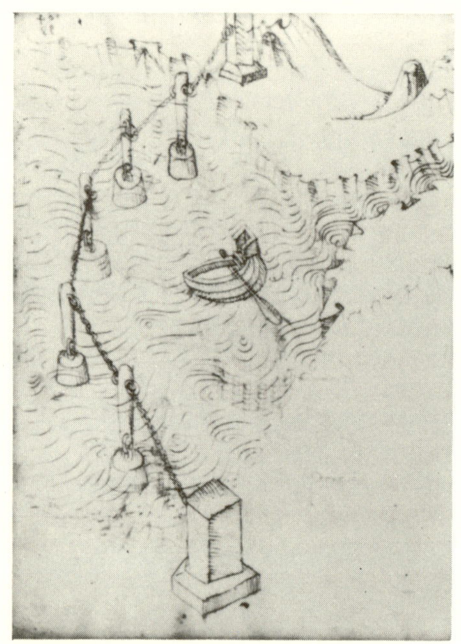

110. Harbor chain (variant) **(Lat. 197, 98r, sequel)**.

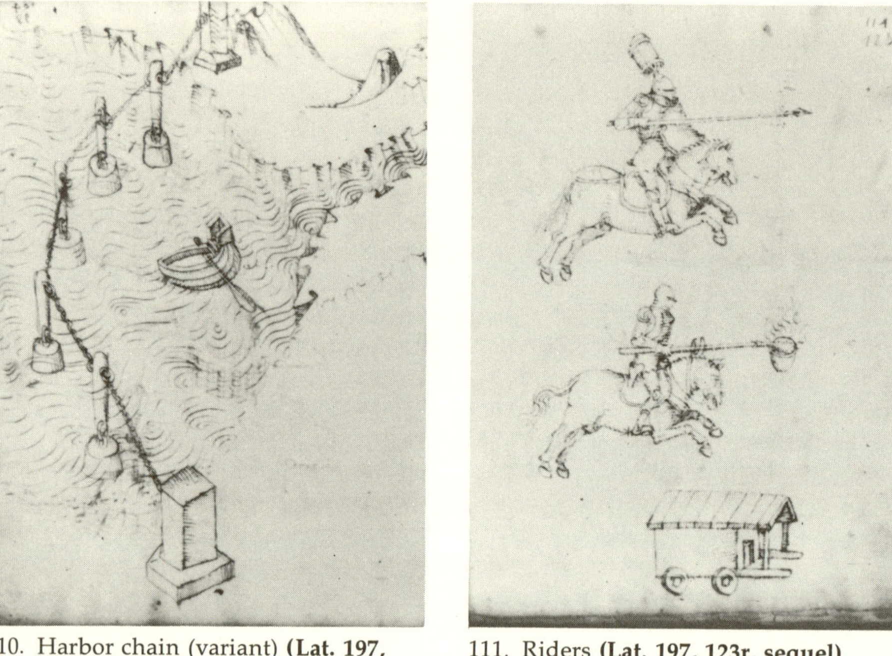

111. Riders **(Lat. 197, 123r, sequel)**.

112. Foot soldiers **(Lat. 197, 136v, sequel)**.

113. Man operating a bucket wheel **(Lat. 197, 25v, separate fascicule)**.

8

Several copybooks have been cited throughout this publication as manuscripts that contain *De ingeneis* and engineering drawings taken from varied sources. Not one of these copybooks is signed or dated, and although attributions have been proposed by earlier writers, no one has given detailed consideration to them or to their sources, one of which is Taccola's *De ingeneis* itself. It is not possible at this point to present a complete history of this remarkable set of codices; however, the following notes may clarify the identities of persons who either had Taccola's autograph work before them and copied it or who may have copied from copies of it. If our observations as to dates and artists are correct, these works were produced throughout the second half of the fifteenth century and beyond, during the times of Francesco di Giorgio, Leonardo da Vinci, and their early followers.

Urb. Lat. 1757 in Rome is an unsigned and undated booklet, the pages of which measure 81 × 59 mm. It is almost 3 cm thick and contains fine parchment leaves, numbered from 1 to 236.[104] Leaves 175 to 218 are lacking, and five pages between folios 45v and 50r are blank or erased. Many drawings in this *codicetto* are combined with legends and texts, generally in Italian, which in many cases amount to simplified but otherwise fairly literal translations of Taccola's Latin into the Tuscan dialect. Some peculiar expressions are rendered identically (for instance, on 15v, the *barche coperte super quas edificantur pontes* of Lat. 197, S 107r). The text is written in a rapid and characteristic hand that remains unchanged throughout the book. Michelini Tocci and others identify it with an "early form" of the calligraphy of Francesco di Giorgio (1439–1501). The illustrations are executed in the same ink and usually with the same pen as the text. On stylistic grounds Papini, Salmi and others have accepted the view of Adolfo Venturi, that it is the work of Francesco, and we believe the work is among his earliest, ca. 1460–1470. The manuscript, like most of the other copybooks, contains none of the drawings of *De machinis*, as distinguished from copies of the Sketches and Sequel in Lat. 197, of which it has many. However, at one point (50v–51r), a chapter title, *De la providentia dela guerra*, is translated from a title in Mariano's *De machinis* (Lat. 7239, 5r: *De provisione guerre*).

---

[104] The entire work is to be published by L. Michelini Tocci. A few folios have been reproduced (Papini, *Francesco di Giorgio*, II, Figs. 301–302; M. Salmi, "Disegni di Francesco di Giorgio nella collezione Chigi-Saracini," *Quaderni dell' Accademia Chigiana*, No. XI, 1947, Figs. 27, 28, 29 (fols. 56v, 71v, 84v–85r). The manuscript was first attributed to Francesco di Giorgio by L. Venturi, "Studi sull' Palazzo Ducale di Urbino," *L'Arte*, 1914, pp. 450–452. The early history of the manuscript is in C. Stornajolo, *Codices urbinates latini*, III, Rome, 1921, p. 678.

The first three scores of folios in the *codicetto* are almost entirely devoted to straightforward copies from Taccola, and here the copying is extremely close; Francesco copies directly from Taccola's autographs. Nothing is added from any other treatise or notebook. Recognizable sequences from Lat. 197 are often preserved.[105] At one point (14r) he writes, *Disegno es(e)so cfr fron[te] 38,* and the *disegno* is a copy of a minute sketch by Mariano in Lat. 197, 38r (now 42r). The contents of the *codicetto* substantially exhaust Lat. 197. By contrast they cover Palat. 766 more selectively and represent only four major drawings. Occasionally the book reflects landscapes by Taccola. Folio 8r reproduces mountains and waterworks from the bottom of Lat. 197, S 107v, and adds a city. Instead of Taccola's terse note *per toleto,* or *Tolleti* as he writes it on I 31v (Pl. 7), it indicates, *A toledo pasa el fiumi da due canne dal monte.* It shows the two pipes: a siphon followed by an inverted siphon. Landscapes are rare, as are figures of men and animals; one receives the impression that a more abstract mind than Taccola's is at work. They are not in a form bearing any strong or obvious resemblance to Francesco's own graceful style in paintings and sculptures.[106]

Beginning near folio 60 or 100, new technical contributions begin to out-

weigh Taccola's work, although copied engine drawings continue to appear. There are new fortress plans (59v, 60r). They show nothing of the bastioned front that part of the literature likes to attribute to Francesco, but they show active work in studying functions of various structures. There are new forms of pumps and siphons (68v, 69v), and there are progressively clarified drawings of Chevaux de Frise (including two on 99v). At one point (101r) there is a change for the worse: Francesco attempts to improve Taccola's recirculation system for mill water (III 33v–34r), but converts it into a fantasy of the perpetual motion type. A preceding page (98r) actually speaks of *continuo moto.* The miller would start the pump by turning a capstan and thereby driving a cogwheel system and horizontal camshaft, and the pump (it was hoped) would then pump water onto a waterwheel to drive this shaft, not only the mill. The fruitless idea reappears frequently in later books.

Urb. Lat. 1757 shows Taccola's riverworks in vastly superior perspective (for instance 109r). It includes a progression of sketches and drawings of gear systems for steering a car or stage wagon (113r–119v). This is followed by a Gear Complex, showing column lifters, warship propulsions, and related mechanisms.[107] Was it the copyist who developed these ideas? We can hardly think that he did the first drafts of mechanical innovations on paper of such minute

---

[105] For instance Lat. 197, fols. 113, 112, 111, 110, 108 recur on fols. 1, 2, 3, 4, 5 of the copybook, in this sequence; fols. 52, 51, 42, 40 recur at fols. 6, 7, 8, 9. There are even longer sequences preserved except for minor breaks; for instance fols. 7, 8, 9, 10, 11, 12, 13, 14, 15 of the autograph recur on fols. 27, 28, 29; 22, 23, 24; 30, 31, 34 of the copybook.

[106] Papini, *Francesco di Giorgio,* II, *passim;* Weller, *Francesco di Giorgio, passim.*

[107] Certain copies of this Gear and Car Complex are described and discussed by Scaglia (1966). This set also occurs in Add. Ms 34113, 113v–131r.

size, even if these leaflets contain a
larger and more varied record of the
Gear Complex than we find elsewhere.
However, actual conclusions must await
the publication of the *codicetto*.

This booklet includes a clock driven by
a compression spring and regulated by
a spring escapement (126r). Instead of
Taccola's pump, mill, and landscape
illustrations, the booklet has a long series
of schematic drawings, dissecting the
original 2-page combinations and pro-
liferating the contents of the 1-page
drawings. We call this part a Mill and
Pump Complex. Its origin is unknown,
but various details in it have forms
characteristic of Taccola's work. Other
details are in new form. Almost all machines are shown in tight fitting wood
frames. Levers and rods are interconnected by slot and roller linkage (142 ff).
Some of the new pumps are of a swing-
ing-piston type (146r, 148r; also 68v, the
latter with cam drive known from Taccola, Lat. 197, folio 134v). All this then
recurs in other engineering manuscripts
discussed here and in the so-called *Trattati* of Francesco di Giorgio.[108] In Urb.
Lat. 1757, the drawings of the Mill and
Pump complex are repeated once more
in minute diagrams, arranged in orderly
tiers and columns on each of a number
of the small pages (163r to 169r). Some

of the final sketches in the booklet (169v
to 174r) are unfinished. The end of the
book (219r to 235) is occupied by the
Marcus Graecus text.

Francesco's abilities as a draftsman
may be noted on every page of Urb.
Lat. 1757. However, he suppressed his
graphic ability as best he could, in the
interest of copying Taccola's work exactly. On and after folio 60, he begins
to introduce his own technical ideas,
often including his own technical errors
or at any rate the ideas and errors of
men other than Taccola. Perhaps he
began, as a young apprentice, to "learn"
mechanics from Mariano's works but
was soon recognized as one who could
clarify these works. Consistently he im-
proves the composition and execution
of Mariano's machine drawings as technical disclosures and interprets them
plausibly; so for instance in working on
pumps and other engines, folios I 60v,
62r, 63 of Mariano (here folios 41r, 33v,
11v, respectively). Perhaps the apprentice then began to produce new prototypes of his own conception, gradually
leading to the development and proliferation of a large Mill and Pump Complex, an equally large Gear and Car
Complex, and new patterns for the showing of fortresses, cannons, Heronian
vases, and other devices. Some of these
may only appear as new prototypes and
may actually be copies of lost works
originated by Taccola or other persons.

In the seventeenth century the *codicetto*
turned up in the vicinity of Urbino.
Francesco may have taken it there when
he went to the court of Duke Federico
in 1477 and may have discarded it when
he was more advanced in his own for-

[108] These are Palat. 767, pp. 60–87; S IV 5,
94r–95v; Add. Ms 34113, 224r–249v; the so-
called *trattati* of Francesco di Giorgio, ed.
Maltese (1967), I, Pls. 62–75, and their later
versions: *ibid.*, II, Pls. 325–331, also S IV 1,
108v, 109r, 119r–120v; "Anonymous" drawings 520A, Uffizi, Florence; and certain drawings in Milan, Modena, and Urbino. Some of
the mill and pump drawings reappear in
printed books: L. Reti, in *Technology and Culture*, IV, 1963, pp. 287–298.

mulations. Before then the contents of the booklet had begun to influence others in Siena, who copied directly or indirectly from it. In 1628, the booklet was listed as part of the library in a ducal villa at Castro Durantis near Urbino. By then the house of the Dukes of Urbino had died out, the region fell to Rome, and the booklet then came to the Vatican Library.

Palat. 767 in Florence is a parchment codex, containing about 264 pages (132 sheets) of somewhat smaller format (28 × 18 cm) than Taccola's autographs.[109] About 95 pages are blank, and others show traces of drawings in draft, indicating that the work is unfinished.

The front page contains the statement: *Del Senatore Carlo de Tommaso Strozzi 1670.* The year is that of the Senator's death. Some one of the Strozzi family may have acquired the book in the Quattrocento.[110] The Senator's library came subsequently into the Florentine Cabinet (later Museum) of Physics and Natural History, now called the Museum of the History of Science. Finally the books came into the Biblioteca Nazionale. It seems that the Cabinet of Physics was classified as part of the Palatine collection. As noted by Gentile (1885) and Jähns (1889), the manuscript contains a copy of Palat. 766.[111] It also contains

material from Books I and II and the Mill, Pump, and Gear Complexes. In all cases the copying is close. The manuscript copies Books III and IV in a form independent of Urb. Lat. 1757 and in much wider selection, while it copies large portions of the other contents in a form that repeats peculiarities of the Urbino *codicetto*. In particular, many deviations from Lat. 197, found in the Urbino booklet, reappear here. The impression arises that Palat. 767 is based on Palat. 766 and Urb. Lat. 1757, not Lat. 197.

The first 237 pages are devoted to drawings of engines and some figures (pages 5, 9, 10, 116),[112] one of the figure drawings (116) being combined with the showing of a lake and surrounding countryside. Except for two brief notes, this part is without text. At the end (pages 238–241 and 250–252) is an Italian translation of some of Taccola's texts in Books III and IV and the opening part of the Marcus Graecus copy from Book II. Blank folios after these translations indicate that full translations were projected. The text, written by a scribe, is finely balanced and the phrasing fluent. The trans-

---

109 Listing of the contents in Gentile, *I codici palatini*, p. 298 f.

110 See p. 31 f. Further details about Strozzi and his library in *Le carte strozziane del R. Archivio di Stato di Firenze. Inventario*, ed. C. Guasti and G. Milanesi, Florence, 1885, pp. v f. See a biography of Carlo di Tommaso di Simone Strozzi in S. Salvini, *Fasti consolari dell'Accademia fiorentina*, Florence, 1717, pp. 461–476.

111 Gentile (*I codici palatini*, p. 298) commented on the existence of similar drawings in Fran-

cesco di Giorgio's *Trattato*. Gille (*Les ingénieurs*, p. 238) is not explicit in making an attribution. He lists the codex under Francesco's name, grouping it with manuscripts known to be copies and with others once assigned to Francesco but removed long ago from his authentic works. The codex has not otherwise been discussed in the modern literature.

112 We indicate the contents partly by footnotes to corresponding chapters in Lat. 197 and Palat. 766 (Books I to IV) and partly by listing the Gear Complex (GC) and Mill Complex (MC) drawings in the Classified Index of Technical Concepts, pp. 222–226.

lation is more literal than in Lat. Urb.
1757. It is not by an engineer, nor do the
drawings betray mechanical ability.
Wherever there is confusion or complexity in the autograph, this copy com-
pounds it. In this way it differs from
Urb. Lat. 1757, where the correlation
between drawing and text is almost perfect and where the technical difficulties
are—with rare exception—resolved rather
than aggravated. The authors of Palat.
767 did not even bother to correlate the
drawings, texts, and vignettes; they
have the vignettes on the wrong pages
consistently, if one can be consistent in
copying erroneously.

In spite of these shortcomings, Palat.
767 is a useful tool for interpreting Taccola's manuscripts and their historic
position. The translations allow us to
check our own reading of the autograph.
The drawings confirm a point that might
otherwise be questioned: no change has
taken place in the constitution of Palat.
766 since the Quattrocento. Even where
Palat. 767 omits a drawing (it omits
most of the two-page scenes), it retains
evidence that the drawing was there.
For instance, it omits the mill and pump
system of III 33v–34r but retains the
she-wolf of 33v (page 35). The Mill and
Pump Complex is added. The impression
arises that a new, improved edition of
De Ingeneis was intended.

The identity of the artist or artists of
Palat. 767 may bring us closer to knowing when, where, and how the *De ingeneis*, with its supplements, was studied. There is no signature, recorded
date, or artist's calligraphy, so there is
no substance to an attribution to Francesco di Giorgio recently proposed by

Gille (1964). In the vast majority of drawings a single, characteristic technique
can be recognized. The artist copies illustrations by applying deft brushstrokes
of ink washes, after making outlines (Pl.
126). Many drawings, especially in the
Mill Complex, are left in the outline
stage of work. In those that are finished,
brushstrokes articulate parts of engines,
human figures, animals, and bodies of
water. The shape of these washes, and
their rhythmic effect, is unmistakably
consistent wherever they are used, which
is to say in the bulk of the manuscript.
This technique occurs in both the Taccola copies and those of the Mill and
Gear Complex. Three divers and a fountain figure (pages 5, 9, 10) are different
in technique, since pen strokes are used
in modeling (Pls. 127, 128). As the ink
tones of these are the same as in the
rest of the codex, perhaps the artist
merely used a different technique in
figure drawing.

The copyist working with a brush makes
slight improvements in the perspective
of wheels and structures, changes the
shapes of wells and fountains, and one
of the boats. He substitutes Taccola's
Gothic fountain figure of III 28r with a
nude youth, the back of his hand pressed
against his hip, standing on an elaborate
candelabra support in the center of the
fountain. His changes in proportions and
postures of the engine operators and
their dress—a very short, neatly pleated
tunic—give them the appearance of
small boys. He represents Taccola's
clothed miller, IV 65r, as a nude youth
and completely alters the pose of a putto,
among the vignettes, by showing him
dancing and using a whip. He trans-

forms the severe rocks of Taccola's landscapes into dainty arabesque patterns.

Almost all these motifs originate in the
paintings of Vecchietta (1412–1480), and
some of them continue in his pupils'
works dating 1475–1485. Vecchietta was
active as a military architect at Sarteano
and Orbetello in 1467–1470.[113] As engineer and artist, he had more than
passing interest in the subjects contained
in this copybook. However, the absolute
lack of engineering originality in this
manuscript indicates that the copywork
is not by Vecchietta himself but by one
of his artist pupils. Francesco di Giorgio,
who may have been Vecchietta's pupil,
may be eliminated from the list of possible artists of this book. Nowhere in the
sedate and static figures of Palat. 767 is
there a trace of the vitality of line and
the qualities of style such as he had
achieved in manuscript illuminations
and paintings as early as 1457–1467.[114]
The codex does have a number of motifs
similar to the work of Neroccio de' Landi
(1447–1500), who was in partnership
with Francesco for a few years until its
termination in 1475.

The pilaster capitals of the fountain
house (34), corresponding to III 31r, are
identically represented by Neroccio in
the *Miracle of the Funeral of St. Bernardino*, ca. 1467–1470.[115] His panels of the
Stories of St. Benedict, ca. 1470–1471, include a number of youths wearing very
short tunics who are very similar to the
small-boy operators of engines in Palat.

767.[116] On the other hand, the same type
appears in narrative panels in Liverpool
and Munich, which are now attributed
to Benvenuto di Giovanni (1436–1518)
and were painted, it seems, soon after
1460.[117] The nude miller and the fountain
figure (21, 103), corresponding respectively to IV 65r and III 28r, develop out
of figures in frescoes by Vecchietta and
recur in Francesco di Giorgio's stucco
relief, *Discordia*, ca. 1475–1485.[118] Vecchietta first shows rock forms in arabesque patterns in his *Assumption* of
1462–1463.[119] In later works he and his
pupils modify these forms, arranging
them in an effect of broken boulders.
That form of rock appears in the lake
scene (Pl. 129) in Palat. 767 (116) as well
as in Neroccio's *Stories of St. Benedict*,
ca. 1470, and in Benvenuto di Giovanni's
Adoration of the Child, ca. 1483.[120] Con-

[116] Coor, *Neroccio*, Figs. 15, 16; Pope-Hennessy,
Sienese Quattrocento, Pl. 84 (*St. Benedict Receiving Totila*). All three panels are illustrated, with an attribution to Francesco, in
Weller, *Francesco di Giorgio*, Figs. 15–17.
[117] Both panels are attributed to Benvenuto di
Giovanni by Pope-Hennessy (*Sienese Quattrocento*, p. 30), where the Liverpool panel of
San Bernardino Preaching is Fig. 17. The panels
have also been attributed to Vecchietta,
among other artists.
[118] Cf. the figures in frescoes of the *Flagellation* and of the *Road to Calvary* by Vecchietta
and his pupils in the Baptistery, Siena (van
Marle, *Development*, XVI, Figs. 126–127). See
the reliefs of *Discordia* in Weller, *Francesco
di Giorgio*, Figs. 54, 55.
[119] Pope-Hennessy, *Sienese Quattrocento*,
Pl. 49. See also detailed photographs in the
Frick Art Reference Library.
[120] See the *Stories of St. Benedict* in Weller,
Francesco di Giorgio, Figs. 16, 17; Coor, *Neroccio*, Figs. 14, 16. Benvenuto di Giovanni's
Adoration in the Van Gelder collection, Brussels (van Marle, *Development*, XVI, Fig. 235),
and other photographs in the Frick Art Reference Library.

[113] G. Vigni, *Lorenzo di Pietro detto Il Vecchietta*, Florence, 1937, pp. 66.
[114] Cf. Weller, *Francesco di Giorgio*, Figs. 1–12.
[115] G. Coor, *Neroccio de' Landi*, Princeton,
1959, Fig. 8.

sidering the lake scene as a whole, it seems to have close connection with a painting on a book cover by Guidoccio Cozzarelli (1435–1517), *The Madonna Guiding Into Port the Ship of the Sienese Comune,* ca. 1475.[121] Both scenes show a large body of water, and a walled citadel with similar architecture—a domed church, a palace, and towers.

These thematic and stylistic parallels make it likely that this manuscript ought to be dated in 1467–1470 at the earliest or at the latest, 1475. If it is not the joint work of Neroccio, Benvenuto, and Guidoccio, perhaps it is an early work of Neroccio, attributable to the indicated time. The qualitative effect of ink washes seems to reflect a peculiarity of his technique in tempera painting. As a painter, he achieves illusions of delicate flesh tones by laying on colors in thin, superimposed layers without varnishes.[122] He is not the only artist to use such media, but his developed use of the technique is fairly characteristic and may well be related to the ink-wash technique of the drawings in Palat. 767.[123] A drawing of the *Triumph of Bacchus,* formerly attributed to Francesco di Giorgio and now given to Neroccio, uses ink washes similar to those of Palat. 767. It is true that the figures are more elegant and more classic than in this copybook.

At least one drawing in Palat. 767 is clearly the work of Guidoccio Cozzarelli. This is a pen drawing (Pl. 120) of a female figure designed as a fountain figure (5). She wears Grecian dress, which billows about the body and is drawn with copious pen work. Similar types of figures appear in Sienese painting and sculpture, the earliest of which is an angel in Francesco di Giorgio's *Coronation,* ca. 1471.[124] What differentiates the fountain figure from Francesco's angel is her face, if not the poor drawing of anatomy. Her broad cheeks, the hard, inflexible line tapering to a small chin, and the large nose and mouth are all the characteristics of many women painted by Cozzarelli.[125] Perhaps we may attribute to him also the nudes using swimming equipment (9, 10), since the pen technique is the same, and at least one of them shows the poor anatomical articulation of Cozzarelli's figure. However, the swimmers are not without similarity to the figurative types of Benvenuto di Giovanni. His paintings of male saints traditionally shown as nearly nude figures resemble these figures in outline and modeling and also in facial features and fluffy hair style.[126]

---

121 In the Archivio di Stato, Siena; photographs are in the Frick Art Reference Library. The lake scene in Palat. 767 and its copy in Ital. Z 86 seems to portray Lago di Nemi, seen from the via Appia. A youth standing on shore points to a ribbed object sunk in the lake. He and the horseman to whom he indicates the sunken treasure are characteristic of figures painted by several Sienese masters, including Neroccio de' Landi.

122 Coor, *Neroccio,* p. 15.

123 Coor, *Neroccio,* Fig. 141, with an attribution to Francesco di Giorgio.

124 See the angel standing beside Christ (Weller, *Francesco di Giorgio,* Fig. 31) and the bronze candle-holding Angels in the Duomo, ca. 1495–1497, in *ibid.,* Figs. 86, 87.

125 Among many pictures by Guidoccio Cozzarelli, see the *Annunciation and Flight into Egypt,* National Gallery, Washington (Pope-Hennessy, *Sienese Quattrocento,* Pls. 66–68).

126 Cf. these features in the figure of *St. Jerome* (Shapley, *Paintings from the Samuel H. Kress Collection,* Fig. 441) and the quasinude youths in the panel, *Christ Carrying the Cross,* Fig. 429.

If it was not the circle of Vecchietta that brought the copyist of Urb. Lat. 1757 into contact with the copyists of Palat. 767, the Cozzarelli family may have been the intermediaries. Giacomo di Bartolomeo di Marco Cozzarelli (1453–1515) was Francesco's pupil, and he worked at Urbino for ten years. Giovanni Cozzarelli, father of Guidoccio di Giovanni Cozzarelli, was one of Francesco's assistants in 1472–1475 when Francesco, with others, worked on the *bottini* (*butini*) of Siena, and was associated with him in bridge construction at Macerata in 1487.[127] Francesco's *codicetto,* or some copy of it, seems to be the basis for Neroccio's work in Palat. 767, so far as the material from Books I and II is concerned.

Additional influences are also possible. For instance, Palat. 767 contains several drawings of bombards with ornamentation (162, 163), one of them inscribed *Opus Dionisio* and one having the Orsini coat of arms. Dionisio of Viterbo, according to documents of 1469 to 1477, worked in metal.[128] He made a famous clock, iron grillwork, metal fixtures for bookcovers, and hydraulic devices. In 1481, he went to Venice for work in connection with its canals. The bombard designs probably represent his work but are not necessarily his drawings. However, the record once more substantiates the pro-

posed dating of the codex in the late 1460s or early 1470s.

S IV 5 in Siena comprises some 94 folios of large format, measuring 25.5 × 35.5 cm. Originally it comprised at least 143 sheets, according to librarians' notes in it.[129] The drawing technique and the handwriting throughout the codex make it fairly certain that the bulk of it is the work of a single artist-author, who copies and develops material from different sources. The ink used is very uniform, as are certain color washes applied to the more finished drawings.

The book now contains only 10 drawings copied from Taccola's Books III and IV.[130] About four to five dozen drawings are from Book I, in the modified form that is also reflected by other copybooks. Several scores are from the Gear Complex, while only four drawings are from the Mill Complex. In addition there are cannon drawings, showing some of the more advanced Renaissance mountings (1r to 4v), and two of the cannon tubes are somewhat similar to that attributed to Dionisio in Palat. 767. The author also shows a long series of engines for naval war (56r to 75v), related distantly to Taccola's illustrations but with modified and elaborately developed accessories and rigging; most of these versions are page-filling pictures with engines curiously overloading the ships. These are likely to be copied from a prototype now

---

[127] Weller, *Francesco di Giorgio,* pp. 5 f, 20, 344, 361, where there are also a few facts about Giacomo di Bartolomeo. Other data about these engineer-architects in Thieme-Becker, *Künstler Lexikon,* VIII, 1913, p. 37 f.
[128] Milanesi, *Documenti,* II, p. 361. E. Morpurgo, *Dizionario degli orologiai italiani,* Rome (Clessidra), 1950, s.v. Dionisio di Cecco da Viterbo, pp. 58 f.

[129] The paper has a variety of watermarks ranging in date from ca. 1350 to ca. 1450.
[130] The Brunelleschian hoist machines (84v, 93v, 97v), other cranes (11r, 11v, 85r, 97v), two trebuchets (25v, 29r), and roof beams (51r).

unknown.[131] They show keen interest in technology but little mechanical understanding. The codex ends with surveying scenes and discussions (99r to 108r), related to the first page in Taccola's Book III but sufficiently different to suggest that it is probably copied from some common source, which also reappear in illustrations of a 1568 edition of Alberti's *Ludi matematici* (written ca. 1450).[132] In these scenes the surveyor is indicated by a small portrait bust, and different features are used for each head. The entire copybook indicates that the artist reproduced technical works by Taccola and amplified them with other ideas. The artist, like other copyists, seems unaware of the *De machinis*.

Notes placed in the codex by Del Rosso and Romagnoli, sometime before 1835, state that this work, found in Pisa a short time before, is by Francesco di Giorgio.[133] Promis (1841) concurred, calling it the "Codice di Cecco." The attribution was rightly rejected, although on various grounds and to different extent, by Mancini (1917) and Salmi (1947). Actually, none of the drawings in S IV 5 shows Francesco's confident mastery of mechanics. Clearly the handwriting is not his, and the human figures, which are very few, do not show his elegant rendering. The artist's conception of

landscape is the only major evidence available for his identification. In all of the numerous landscapes from Books I to IV copied here, Taccola's illustration of earth at the edge of water is reduced to fine lines breaking in zigzag courses and producing an effect like a thin sheet of ice. Taccola's rock crags are also given sharp edges. They are modeled by crossing pen strokes. This handling of edges as thin ledges and laminations is characteristic of Benvenuto di Giovanni, who shows them in his *Adoration of the Shepherds,* dated 1470, and in later work.[134] Since we are unaware of other artists who show these characteristics quite so clearly, we suggest, although only provisionally, that the codex may be attributed to Benvenuto di Giovanni. He was another pupil of Vecchietta, a few years older than Francesco di Giorgio. The date of S IV 5 may be approximately the same as that of Palat. 767.

Add. 34113 in the British Museum, London, comprises 261 sheets of paper, measuring 20 × 28 cm and marked with a ladder watermark, probably of the third quarter of the fifteenth century.[135] Some initial pages (9–14) and some end pages (254–261) are of thicker or thinner material, respectively. Otherwise the set appears homogeneous.[136] On most pages,

[131] Some of them are also seen in Add. Ms 34113, 177 ff.
[132] Alberti's text in translation and the illustrations of "Delle piacevolezze delle matematiche," in *Opuscoli morali, tradotte & parte corretti da M. Cosimo Bartoli*, Venice, 1568, pp. 224–253. Regarding the date of writing, see Mancini, *Vita*, p. 281.
[133] See also L. Ilari, *La biblioteca publica di Siena*, Siena, 1848, p. 119.

[134] Cf. Van Marle, *Development*, XVI, Fig. 226.
[135] Briquet, *Les filigranes*, I, col. 344, dated 1450–1474. Also see note 6, p. 26.
[136] It was, however, taken apart and rebound. This is clear from the fact that Van Marle (*Development*, XVI, Fig. 150) presented a photograph of two pages of the codex, showing them as facing one another, while actually the left half of his illustration is fo. 139v of Add. Ms 34113 and the right half is another verso, 110v of this codex. Incidentally he at-

authentic and new paginations are seen.
We use the new pagination.[137] The manuscript has been provisionally attributed
to Francesco di Giorgio.[138] It is unique
among the copybooks, as it contains in
a single volume substantially the entire
technical work of Taccola. It omits from
Books III and IV only such items as his
vignettes, portraits, and the signatures
of his cross-references to Books I and II.
It also contains drawings from the Mill
Complex and other Complexes, a fairly
complete set of drawings from Books I
and II and their interspersed Addenda,
and some Sketches and drawings from
the Sequel and Separate Fascicule. There
is also a much more limited selection of
notes from *De machinis*. The copying is
very close. A few engines, accompanied
by texts that are strongly reminiscent of
Taccola's prose,[139] may reflect lost Taccola drawings, perhaps found in an
archetype known to this artist. An interesting question raised by this copy is
whether the Mill, Pump, and Gear Complexes were assumed to be of the same
authorship as the *De ingeneis*. Another
interesting question is why the *De ma-*

chinis is represented in such small part.
Was it recognized that it, in substance,
repeats the central technical ideas of the
De ingeneis, adding a mass of ancient
war engines thereto?

The bulk of the contents appears to be
copied by a single person. There are
some variations of his calligraphy and
drawing technique, but they are slight
and are likely to be due only to the use
of different utensils and perhaps due to
other momentary influences. The writing appears to be of the fifteenth century. The text is a close, Italian translation of Taccola's Latin, independent of
Urb. Lat. 1757 and Palat. 767.[140] It refers
to Maestro Francesco three times (17v,
129r, 194v), designating him at least
once as *Mº Fra[n]cº di Giorgio da Siena*.
The words *da Siena* indicate that the
copybook was produced in a town other
than Siena and where work by Francesco
was known—perhaps in Urbino or in
another of the cities where documents
relating to Francesco di Giorgio cite
him as Francesco da Siena.[141] It appears
probable that the book dates after 1475.[142]
In one of the references to Francesco
di Giorgio, one idea is credited to *Mº*
dion[i]zi (17v) whose name is also mentioned in Palat. 767, and it is connected
with the work by Francesco. One text
(129r f) that cites Francesco's name is a
translation from *De machinis*,[143] while

---

tributed the two pages to Francesco di Giorgio
and stated that the codex is in Siena (actually,
London) and particularly in a "Municipal
Library" there (nonexistent). One may wonder
about the merit of attributions of this type.

[137] There are seventeen blank pages: fols.
16–20, 24, 28, and 29, all recto and verso, and
25 recto.

[138] A. E. Popham and P. Pouncey, *Italian Drawings in the Department of Prints and Drawings
in the British Museum, XIV–XV Century*, London, 1950, p. 38.

[139] See for example 98v, *Quando vole il archi-
tetore votiare iloto de chanali . . .* ; 134r, *In
diverssi modi si fano moltti lacci da ruvinare
sichome vedi nela presentte fighura . . .* ; 141v,
A provedere che li nemici entratti nella citta . . . .

[140] Reference to the Lion and Dragon Books
are retained (24v, 36v). The copyist cites a
"Falcon Book" (41v, 47v), mistaking it for
Taccola's falcon page.

[141] See the documents in Weller, *Francesco di
Giorgio*, pp. 354, 356, 369.

[142] Francesco was not a master before 1464; he
took a pupil only in 1468 and did not leave
Siena before 1475–1477.

[143] The title of this text, one of a number that

two others (17v, 194v) concern the composition for the *lima sorda*, a file represented without annotation in Taccola's Book I, 18v. Here we may have a beginning of the influence that caused Francesco's fame to overshadow the name of the Sienese Archimedes. Clearly, Add. 34113 is of the time when Taccola's total work still was intact as a group of textbooks but was about to be replaced by Francesco's *Trattato*. There is no reflection of the *Trattato* itself in this copybook. We think it was completed before 1500.

Some material in this copybook may be identified as a collection or group of books originally owned by Taccola. Directly preceding the transcription (18v–20v) of Mariano's copy of Marcus Graecus, the copybook includes a discussion (9r ff) of well digging, ore mining, smelting, and related procedures. Directly after the Marcus Graecus transcription, about five pages are blank, indicating that further material was to be added but that the prototypes could not be made available. They may have been in the parchment codex later owned by Vannoccio Biringuccio, since the *Pirotechnia* by this writer combines the Book of Fire with teachings about water and metal.[144]

Internal evidence shows a few facts about the history of Add. 34113. Near the end, in a part composed of thin paper, later Florentine artists added some pages of notes and sketches regarding work in SS Annunziata in Flor-

ence (252v) and in the Certosa of Pisa (253v). The apparent relationship of Add. 34113 to the architectural-engineering circle of Florence reappears in statements in two book notes that mention Lorenzo della Golpaia, Maestro Bernardio, and Matteo Nigetti.[145] Although these men were not necessarily successive owners of the codex, it is at least clear that its history is linked with Florentine developments. The history of the manuscript from the middle of the seventeenth to the middle of the nineteenth century is obscure. Somehow the book was acquired by Sir J. C. Robinson, who sold it to the British Museum in 1891.[146]

Ital. Z 86 in Venice illustrates the fact that later copies were made from the earlier copies. With a single exception (58r, showing III 33v–34r; the equivalent earlier copy seems to be lost), there are no drawings in this manuscript that are not found in Palat. 767, while there are several in Palat. 767 that are not found here. About one third of the drawings here are from *De ingeneis* as copied in the Palatine manuscript. They include about two dozen from Books III and IV, complete with the Palat. 767 version (and

---

can be traced to the prefatory texts of *De machinis*, reads, *Della providentia della ghuerra sicondo M° fra[nces]c° da Siena detto di sopra.*

144 C. S. Smith and M. Gnudi, *The Pirotechnia of Biringuccio*, New York, 1942, pp. 13–85, 338–357, 436–439.

145 Lorenzo della Golpaja (Volpaja), 1446–1512, clockmaker and author of a codex copied by his son Benvenuto di Lorenzo in Cod. Ital. IV, 41 (5363). On the codex of Benvenuto di Lorenzo della Golpaja, see C. Pedretti, *Studi Vinciani* (Travaux d'Humanisme et Renaissance, XXVII), Geneva, 1957, pp. 23–33, 99–106. Maestro Bernardio is Bernardo Buontalenti, 1537–1608, master of Matteo Nigetti (d. 1649).

146 *Catalogue of Additions to the Manuscripts in the British Museum, 1888–1893*, Order of the Trustees, 1894, XV, p. 820. Sir J. C. Robinson (1824–1913) was an honorary member of the Academy of Fine Arts of St. Luke in Rome, Florence, and Bologna (*Who Was Who, 1897–1916*, London, 1919, p. 605).

misapplication) of the vignettes. Another third is from the Mill and Pump Complex, and the balance is from the Gear Complex and a few other additions appearing in Palat. 767. The copying from Palat. 767 is particularly apparent in drawings such as the lake scene, the bombards, the divers, and the engine operators. For instance, the main artist of Palat. 767 had substituted nude figures for Taccola's miller, crank operator, and a recoverer of sunken goods; they and details of their execution recur in this new manuscript. The copying includes an attempt to model with brushstrokes, but the execution is noticeably inferior to that of the earlier copy. The technique indicates that it was probably executed in the Quattrocento, perhaps soon after Palat. 767. There are substantially no texts in Ital. Z 86, and there is neither a date nor signature, as is usual in these copybooks. It was part of a set of books bequeathed to the Biblioteca Marciana by Jacopo Contarini of Venice in a testament of 1595, fully executed only in 1713.[147] Morelli (1776) drew attention to the manuscript. Promis (1841) attributed it, like most other technical manuscripts, to Francesco di Giorgio.[148] Canestrini (1940, 1946) recognized it as traceable to Taccola, although he was unaware of its direct source, Palat. 767, and its principal, ultimate source, Palat. 766.

The copying of information from *De ingeneis* continued for a long time. One of those who did such work was the Sienese architect Bartolommeo Neroni, called Il Riccio (ca. 1500–1572), the reputed author of Codex S IV 6 in the Biblioteca Comunale, Siena.[149] The contents are heterogeneous. One page (34r) is unique and is likely to have been of more than common, antiquarian interest already in Neroni's time, as it reflects an architectural concept of Brunelleschi in a form for which no equivalent record is known to exist. The drawing is made with ruled lines forming the diagram of a housefront. It shows three stories in elevation, each proportionally subdivided with three openings. The openings or windows in the upper stories are roundheaded, as is the central portal on the ground floor. The lower part of the page is filled with a note in Italian, by an unknown hand of the Quattrocento that gives instruction about proportioning and constructing such a facade.[150]

[147] The Contarini bookplate is on the verso of the first folio (numbered 13), and the record of 1713 appears on the inside cover.

[148] Promis' error in citing the manuscript is corrected by C. Frati and A. Segarizzi, *Catalogo dei codici marciani italiani*, II, Modena, 1911, pp. 5 f.

[149] The attribution was made by G. Milanesi, in an unpublished note in the codex. An earlier note in the codex said that some folios, such as 26 (?) and 34, seem to be by Francesco di Giorgio. On this question, see also Ilari, *La biblioteca publica di Siena*, p. 119. Actually, fo. 26 is part of a set clearly shown by calligraphy to be much later than Francesco di Giorgio, as is 34v, and their drawing techniques are not similar to his. Folio 34r is in a technique and calligraphy much older than Francesco's time; see the text that follows.

[150] Nota se voi fare una faccia duno chasamento in prima piglia il suo diametro e cioe la sua largheza / e partirai le porti che a tte pare che si conveghino sappi che ogni chasamento / vuole una porta / ho tre ho cinque o sei o / nove o undici e chosi farai di sopra le finestre conferire vacuo sopra vacuo e pieno sopra pieno / notificandoti che tanto vuole essere el pieno quanto il voto e ogni largheza di finestra o porti vuole esser alta dua tanti quanto le largha e cioe due quadri ancho farai diminuire la groseza del muro di dieci

Six pages (31r–33v), directly preceding this one, are devoid of notes but show machines from the modified form of Book I and II and the Gear Complex. By the manner in which ink washes are applied to these line drawings, the artist of this part of the manuscript may be the same one who did the greater part of Palat. 767. If Neroccio de' Landi produced the parchment edition, Palat. 767, he also seems to have made the paper versions on these pages. S IV 6 also includes Quattrocento drawings, without notes, which show architectural entablatures, capitals, column bases, ornaments, and geometric constructions or vase forms (1–30, 54–66). Still farther on there is about a score of sheets (34v–53v) that contain belated copies from *De ingeneis* and the Gear and Mill Complexes. As to these the attribution to Neroni is plausible. In this late set the Brunelleschian machine with reversing clutch appears twice, in forms derived from sources other than Taccola.[151] These consist of a perspective view (39v) very similar to the drawings of the same machine made by Giuliano da Sangallo and Leonardo da Vinci and orthographic projections of the major component parts (49r). We

may safely conclude that someone like Neroccio both collected and copied drawings of Brunelleschi's and Taccola's time and circle and that men like Sangallo, Leonardo, and Neroni later on both collected and copied drawings of this kind. They gradually developed the graphic methods for illustrating the machines and, more gradually, no doubt developed the machine concepts themselves.

Another collection of similar type has been attributed to a still more famous architect, Antonio da Sangallo the Younger (1483-1556).[152] The Uffizi drawings in the Gabinetto dei Disegni e Stampe contains a complete set of drawings, closely and competently copied from a modified copy of Books I and II and the Mill and Gear Complexes, together with a few sketches of similar content, possibly by other draftsmen. The paper of these drawings shows that a great part of the contents at one time was bound in book form. The drawings preserve Taccola's composition, viewpoint, and details in their entirety but clarify graphically what parts are of wood or of metal, and other details of this kind. In the highly diagrammatic Gear Complex the drawings also clarify proportions of machine parts and forms of elements.

There are also collections that concentrate entirely on the Mill and Pump or Gear and Car Complexes, and wherein even modified copies from Books I and II disappear from sight except for a very few drawings. There are several note books of this type: Codex Magliabecchianus XVIII, 4 (II, III 314), and Codex

---

braccia in dieci braccia secondo assuffitientia di tale edifitio e questo lassarai alle inposte delle volte overo de palchi."
[151] (Fo. 49r): A ser Francisco. Questi pezi di di modelli sono elluidine del disongnio in nella carta scontro come sieno i sui profili el pezo sengniato A e quello delle due Ruote ritto li due altre sono per lle Ruote sengniate B e C. When referring to a pertinent design on a *carta scontro*, the writer seems to refer to the page that now is 39v. The *profili* are views taken almost purely in orthogonal projection, while the other design is in freehand perspective. The housefront on 34r is purely in front view.

[152] G. Giovannoni, *Antonio da San Gallo il Giovane*, I, Rome, 1959, pp. 407 ff.

Magliabecchianus XVIII, V, 2 (*Disegni di machine*), in the Biblioteca Nazionale, Florence; Codex N 191, Biblioteca Ambrosiana, Milan. Additional copies of *De ingeneis* are suggested by various writers but not thus far located. Such a copy may be involved when Vannoccio Biringuccio states that he owned a parchment booklet on incendiaries. More recently, "another copy" of a work by Mariano was mentioned by Laborde (1924) as being in Vienna. It is actually one of the many versions of Kyeser's *Bellifortis.* Our own experience in reviewing manuscripts of this kind leads us to think that still further reflections of Taccola's work may be found.

114. Gualtieri di Giovanni (?), *St. Honofrius*, detail of the Triptych No. 140, Siena, Pinacoteca. (Photo: courtesy of Alinari)

115. Domenico dei Cori, *Goliath*. Siena, Duomo, pavement. (Photo: Anderson)

116. Domenico dei Cori, *Virgin Annunciate*. Museo d'arte sacra, Montalcino. (Photo: Grassi)

118. Domenico dei Cori, *Adoration of the Shepherds*. Palazzo Pubblico Chapel, Siena. (Photo: Grassi)

119. Lorenzo Monaco, *Legend of St. Honofrius*. Accademia, Florence. (Photo: Anderson)

120. Giovanni di Paolo, *The Raising of Lazarus*. Walters Art Gallery, Baltimore. (Photo, courtesy of the Walters Art Gallery)

117. Simone Martini, *Guidoriccio da Fogliano*. Palazzo Pubblico, Siena. (Photo: Grassi)

121. Giovanni di Paolo, *Flight into Egypt.*
Pinacoteca, Siena. (Photo: Anderson)

122. Sassetta, *Temptation of St. Anthony,*
detail. Robert Lehman Collection, New
York. (Photo: courtesy of R. Lehman)

123. Sienese School ca. 1440, *Adoration of the Shepherds,* detail. El Paso Museum of Art, El Paso, Texas. (Photo: courtesy of the Museum)

124. Giovannino dei Grassi, Sketchbook, fol. 5. Museo Civico, Bergamo, (Photo: Alinari)

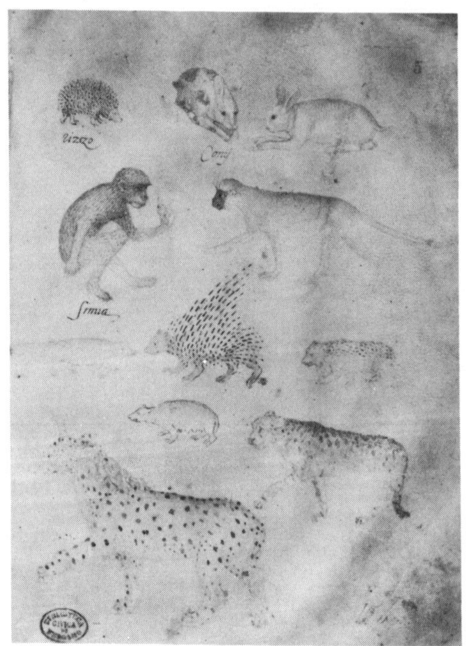

125. Giovanni di Paolo, *Christ in the Garden*. Pinacoteca, Vatican. (Photo: Anderson)

126. Palat. 767, p. 53: Treasure-hunting tools. Biblioteca Nazionale, Florence. (Photo: G. B. Pineider)

127. Palat. 767, p. 10: Swimmers. Biblioteca Nazionale, Florence. (Photo: G. B. Pineider)

128. Palat. 767, p. 5: Fountain figure.
Biblioteca Nazionale, Florence. (Photo:
G. B. Pineider)

129. Palat. 767, p. 116: Lake scene.
Biblioteca Nazionale, Florence. (Photo:
G. B. Pineider)

CONCLUSIONS

9 Mariano Taccola, also known as *Ser* Mariano, friar Mariano, *elegantissimus Marianus,* and the Sienese Archimedes, resided quietly in Siena most of his days, producing secretarial reports, sculptures, portraits, drawings of saints, engineering drawings, and engineering proposals. He gained the friendship of Brunelleschi and other artists and the esteem of political leaders. A German Emperor gave him titles of nobility for his *De ingeneis,* and some four-hundred years later a French Emperor was among those who gave recognition to rediscovered copies of his *De machinis.* During the interval of four centuries the works remained unpublished, but they must have been read and considered for at least two decades after Mariano's death. They were copied by Francesco di Giorgio and many of his contemporaries and followers, down to the times of those who built and debated the cupola of St. Peter's in Rome, the structure that in due course gave rise to the first beginnings of modern civil engineering.[153] Even then, fragments of Taccola's books were incorporated in portfolios of artists and *architettori.*

By then the manuscripts had begun to find their way to the archives where they are now preserved, perhaps with the help of the Order of San Jacopo and surely with the help of Widmanstetter, Strozzi, and keepers of various archives. The codices then emerged from these archives, beginning with the rediscovery of *De machinis* by Carpentier (1766) and Morelli (1776) in Paris and Venice. Shortly thereafter the technical notes by Leonardo were rediscovered by Giambattista Venturi (1797), and it became clear at once that the earlier *De machinis* contained the key to an understanding of such notes.[154] Napoleon III and Favé (1846–1871) and other early writers then discussed detail problems in Mariano's works as represented by the single book then known. Their discussions of Taccola[155] and their technical discussions of Leonardo and others were limited for a number of decades to the rediscovered military concepts. These of course were prominent in Taccola's work, and at least equally prominent in the minds of such men as Napoleon, but they are not as dominant as was assumed.

That the interests of the Sienese master were of broader scope became clear by the researches of Romagnoli, Milanesi, and Gentile, who reported certain facts in the years between 1830 and 1885. Near the end of the nineteenth century, Jähns and Berthelot still were interested in Taccola for no other reason than his significance for military technology, but they became aware of his work of civic nature and drew attention to this work.

---

[153] Straub, *A History of Civil Engineering,* pp. 105 ff.

[154] J. B. Venturi, *Essai sur les ouvrages physico-mathématiques de Léonard de Vinci,* Paris, 1797. See the end of Venturi's final "Notices plus détaillées." The work is reprinted in G. B. de Toni, *Giambattista Venturi e la sua opera Vinciana,* Rome, 1924.
[155] Well summarized by Thorndike (1955).

Since their time and that of Beck (1899) the view which saw Taccola only as a teacher or inventor of war engines has become obsolete and untenable. The histories of technology now consistently show his civic work as well as his military proposals, as can be noted readily by review of the well-known works by Feldhaus, Uccelli, Forti, Singer, White, Daumas, and Gille. However, even those works have given too little attention thus far to the fact that Taccola's personality is expressed in drawings of figures and landscapes as well as hydraulic and mechanical machines and military weapons. To Michelini Tocci (1962), it is less than obvious from the poorly preserved notebook, Lat. 197, that Taccola was an artist. The point becomes obvious only from the more pristine, Florentine part of his *De ingeneis*, Palat. 766.

In order to allow a review of this matter, this publication includes illustrative parts of Lat. 197 as well as Palat. 766. The combined documents, also including *De machinis*, Lat. 28800, show Mariano Taccola as civic official, builder, teacher, and artist, and as able to grow in several of these capacities.

In fact Taccola was so versatile, and some parts of his total work are still so poorly explored, that many problems remain for future study. A determined search should be made for his sculptures. His relationships with men such as Pisanello should be clarified. Enigmatic parts of his engineering should also be reviewed, and the origin of the Mill and Pump Complex and the Gear and Car Complexes should be traced. These problematic drawings should be considered for the light they shed on the proliferation and development of the work by later *architettori,* the large group around Francesco di Giorgio. As the study continues, it may become possible to reach a clearer understanding than is possible today of such men as Alberti, Leonardo, and Biringuccio. From the evidence at hand, Taccola emerges as a truly creative builder of thought, even if he may not be compared with such giants as Brunelleschi and Leonardo. Surely he was an inspired teacher and an expressive artist who produced work that is significant in the trends of Sienese art and in world technology. His *De ingeneis* is one of the important works of modern times and should be of interest to present and future readers.

SELECTED BIBLIOGRAPHY

Aschbach, J. von
Geschichte Kaiser Sigismunds, Hamburg, 1838–1845.

Bargagli-Petrucci, F.
Le fonti di Siena e i loro aquedotti, Florence, 1906.

Beck, Theodor
Beiträge zur Geschichte des Maschinenbaues, Berlin, 1899 and 1900.

Berthelot, M.
"Pour l'histoire des arts mécaniques et de l'artillerie vers la fin du moyen âge," *Annales de chimie et de physique*, ser. 6, XXIV, 1891, pp. 433–521.

Berthelot M.
"Histoire des machines de guerre et des arts mécaniques au moyen âge. Le livre d'un ingénieur militaire à la fin du XIVe siècle," *Annales de chimie et de physique*, ser. 7, XIX, 1900, pp. 289–420.

Berthelot, M.
La chimie au moyen âge, I, Paris, 1893.

Berthelot, M.
"Histoire des corps explosifs," *Journal des savants*, 1895, pp. 687–702.

Berthelot, M.
"Les manuscrits de Léonard de Vinci et les machines de guerre," *Journal des savants*, 1902, pp. 116–120.

Biringuccio, V.
The Pirotechnia of Biringuccio, ed. C. S. Smith and M. Gnudi, New York, 1942. M.I.T. Press Paperback, 1966.

Brandi, C.
Giovanni di Paolo, Florence, 1947.

Brandi, C.
Quattrocentisti senesi, Milan, 1949.

Canestrini, G.
Leonardo costruttore di machine et di veicoli, Milan, Rome, 1939.

Canestrini, G.
Arte militare meccanica medievale, Milan, 1946.

Carli, E.
Scultura lignea senese, Florence, 1951.

Carpentier, D. P.
Glossarium novum ad scriptores medii aevi, Paris, 1766. (Supplement to C. D. Du Cange, *Glossarium ad scriptores mediae et infimae latinitatis*, Venice, 1678).

Daremberg-Saglio
Dictionnaire des antiquités grecques et romaines, Paris, 1873–1919.

Daumas, M. (Ed.)
Histoire générale des techniques, Paris, 1962–1968.

Drachmann, A. G.
The Mechanical Technology of Greek and Roman Antiquity, Copenhagen, Madison, London, 1963.

Feldhaus, F. M.
Lexicon der Erfindungen, Berlin, 1904.

Feldhaus, F. M.
Die Technik der Vorzeit, der geschichtlichen Zeit und der Naturvölker, Leipzig-Berlin, 1914.

Feldhaus, F. M.
Ruhmesblätter der Technik, Leipzig, 1910, and 1925–1926.

Feldhaus, F. M.
Die Technik der Antike und des Mittelalters, Potsdam, 1931.

Feldhaus, F. M.
Die Maschine im Leben der Völker bis zur Renaissance, Berlin, 1954.

Forti, U.
Storia della tecnica in Italia, Florence, 1948.

Forti, U.
Storia della tecnica dal medioevo al rinascimento, Florence, 1957.

Frontinus, Sextus Julius,
De aquis urbis Romae, ed. C. Herschel, Boston, 1899.

Gentile, L.
I codici palatini della Reale Biblioteca Nazionale Centrale di Firenze, II, Rome, 1885.

Gille, B.
Les ingénieurs de la renaissance, Paris, 1964. English translation, The M.I.T. Press, 1966.

Halm, C., and Laubmann, G.
Catalogus codicum latinorum bibliothecae regiae Monacensis, I, 1, Munich, 1868.

Halm, C., and Laubmann, G.
Editio altera, Munich, 1892–1894.

Hardt, Ignaz
Catalogue of Munich "Hofbibliothek," in manuscript, *Cod. Bav. Lat. 65*, Vol. I, 1806.

Hefele, C. J. von
Conciliengeschichte, Freiburg, 1873–1890.

Heron of Alexandria
Opera quae supersunt, ed. W. Schmidt et al., Leipzig, 1899–1903.

Hoffman, D. L.
"Pioneer Caisson Building," *Journal of the Society of Architectural Historians*, XXV, 1966, pp. 68–71.

Jähns, M.
Geschichte der Kriegswissenschaften, I, Munich, Leipzig, 1889.

Kyeser, Konrad
Bellifortis (ca. 1400–1405), Cod. phil. 63, Göttingen. Published in large part by M. Berthelot, "Histoire des machines de guerre . . . ," 1900, *supra*.

Kyeser, Konrad
Bellifortis, ed. G. Quarg, Düsseldorf, 1967.

Laborde, A. de
"Un manuscrit de Marianus Taccola revenu de Constantinople," *Mélanges offerts à M. Gustave Schlumberger*, Paris, 1924, pp. 494–505.

Marcus Graecus
Liber ignium. Published by M. Berthelot, *La chimie au moyen âge*, I, *supra*.

Francesco di Giorgio Martini
Trattati di architettura ingegneria e arte militare, ed. C. Maltese and L. Maltese Degrassi (Classici italiani di scienze tecniche e arti. Trattati di architettura, III), Milan, 1967.

Michelini Tocci, L.
"Disegni e appunti autografi di Francesco di Giorgio in un codice del Taccola," *Scritte di storia dell' arte in onore di Mario Salmi*, II, Rome, 1962, pp. 203–212.

Milanesi, G.
Documenti per la storia dell'arte senese, Siena, 1854–1856.

Monumenta conciliorum generalium (ed. Akademie der Wissenschaften, Vienna), I, Vienna, 1857.

Morelli, J.
Codices manuscripti latini Bibliothecae Nanianae, Venice, 1776.

Napoleon III and Ildefonse Favé
Études sur le passé et l'avenir de l'artillerie, Paris, 1846–1871.

Papini, R.
Francesco di Giorgio architetto, Florence, 1946.

Philon of Byzantium
Pneumatica, vide supra, Heron, ed. W. Schmidt et al., I, pp. 458–489.

Pliny the Elder. Gaius Plinius Secundus, *Naturalis historia*, ed. H. Rackham (Loeb Classical Library), Cambridge, Mass., and London, 1938– .

Prager, F. D.
"Brunelleschi's Patent," *Journal of the Patent Office Society*, XXVIII, 1946, pp. 109–135.

Prager, F. D.
"Brunelleschi's Inventions and the 'Renewal of Roman Masonry Work,'" *Osiris*, IX, 1950, pp. 457–554.

Prager, F. D.
"A Manuscript of Taccola, Quoting Brunelleschi, on Problems of Inventors and Builders," *Proceedings, American Philosophical Society*, CXII, 1968, pp. 131–149.

Prager, F. D., and Gustina Scaglia
Brunelleschi: Studies of His Technology and Inventions, The M.I.T. Press, 1970.

Promis, C.
Della vita e delle opere degl'italiani scrittori di artiglieria, architettura e meccanica militare (in C. Saluzzo, *Trattato di architettura civile e militare di Francesco di Giorgio Martini*), 1841.

Regesta imperii, Part XI, *Die Urkunden Kaiser Sigmunds*, ed. W. Altmann, Innsbruck, 1896–1900.

Reichstagsakten (*Deutsche Reichstagsakten*), ed. Julius Weizsäcker and successors, Munich, 1867– .

Reti, L.
"Francesco di Giorgio Martini's Treatise on Engineering and its Plagiarists," *Technology and Culture*, IV, 1963, pp. 287–298.

Reti, L.
"Helicopters and Whirligigs," *Raccolta Vinciana*, 1964, pp. 331–338.

Reti, L.
Tracce dei progetti perduti di Filippo Brunelleschi nel Codice Atlantico di Leonardo da Vinci (IV Lettura Vinciana, Vinci, Biblioteca Leonardiana), Florence, 1965.

Romagnoli, E.
Biografia cronologica de' bellartisti senesi. Unpublished manuscript (L. II. 4, Vol. IV), Biblioteca Comunale, Siena (ca. 1830), pp. 337–346.

Salmi, M.
"La pittura e la miniatura gotica," *Storia di Milano*, VI, Milan, 1955, pp. 785–855.

Sarton, G.
Introduction to the History of Science, Baltimore, 1927–1948.

Scaglia, G.
"Drawings of Brunelleschi's Mechanical Inventions for the Construction of the Cupola," *Marsyas*, X, 1955–1956, pp. 45–68.

Scaglia, G.
"Drawings of Machines for Architecture from the Early Quattrocento in Italy," *Journal of the Society of Architectural Historians*, XXV, 1966, pp. 90–114.

Scaglia, G.
"An Allegorical Portrait of Emperor Sigismund by Mariano Taccola of Siena," *Journal of the Warburg and Courtauld Institutes*, XXXI, 1968, pp. 428–434.

Schmidt, F.
Geschichte der geodätischen Instruments und Verfahren in Altertum und Mittelalter, Neustadt, 1935.

Thorndike, L.
*A History of Magic and Experimental
Science*, New York, 1923–1941.

Thorndike, L.
Science and Thought in the XVth Century,
New York, 1929 and 1963.

Thorndike, L.
"Marianus Jacobus Taccola." *Archives
internationales d'histoire des sciences*, VIII,
1955, pp. 7–26.

Theophilus
Diversarum artium scedula. Editions by
C. R. Dodwell, London, 1961, and C. S.
Smith, Chicago, 1963.

Tiraboschi, G.
Vetera humiliatorum monumenta, Milan,
1766–1768.

Tiraboschi, G.
Storia della letteratura italiana, Florence,
1805–1823.

Uccelli, A. (Ed.)
*Enciclopedia storica delle scienze e delle
loro applicazioni*, Milan, 1943.

Uccelli, A. (Ed.)
Storia della tecnica, Milan, 1945.

Valois, N.
Le pape et le concile (1418–1450), Paris,
1909.

Vegetius Renatus, Flavius
Epitoma rei militaris. Editio princeps,
Utrecht, 1473.

Venturi, G. B.
*Dell' origine e dei primi progressi delle
odierne artiglierie*, Verona, 1815.

Vitruvius
De architectura libri X, ed. Morris Hicky
Morgan, Cambridge, Mass., 1914 (also
New York, 1960); ed. F. Granger (Loeb
Classical Library), 1929.

White, L., Jr.
Medieval Technology and Social Change,
Oxford, 1962.

Zdekauer, L.
Lo studio di Siena nel rinascimento, Milan,
1894.

Zdekauer, L.
*Il constituto del comune di Siena, dell'
anno 1262*, Milan, 1897.

CLASSIFIED INDEX:
TECHNICAL ILLUSTRATIONS
IN *DE INGENEIS,*
DE MACHINIS, AND
SUPPLEMENTS

Taccola's technical illustrations are listed here under six major classes. The first four of these, as well as their Latin names, are taken from the *tabula* in Taccola's *De machinis* (Spencer Ms 136, 1r, 1v; Lat. 7239, 3r, 3v). The fifth class, Military Engineering, is a condensation of six different classes in this *tabula*. The sixth class, Architectural Drawings, is not used in Taccola's own books. The subdivision of classes under each major heading is a modern one, using Latin terms of Taccola so far as these are available in his writings, along with his occasional use of Italian terms.

The methods of reference as explained at the beginning of Part II are used here and, in addition, the following:
DM
Autograph of *De machinis,* Lat. 28800, Munich (Taccola's pagination)
DMN
Copy of DM in Spencer Ms 136, New York Public Library (Borsatis's pagination)
FG
Sketches on S 130v, probably by Francesco di Giorgio
GC
Gear Complex drawings as shown in Palat. 767

MC
Mill Complex drawings as shown in Palat. 767

The folios where Taccola discusses or illustrates each type of engine are listed here at the left of each column (for example, I 61r or III 43r). The page where each type is first presented herein is listed at the right (for example, 63 ff). Devices not discussed by the authors are marked with an asterisk.

1. CIVIL ENGINEERING
De pontibus et aliis edifitiis super aquas

Box caisson, *cassa,* and equip-   63 ff.
ment for it.
I 61r; III 43r; S 133v; DMN 9r;
details: I 44r, 44v; S 108r, 109r,
110r

Cofferdam, *stechato.*   62 f.
I 74v; II 81v; III 42r; S 121v, 122r;
elements and tools; I 47r, 47v,
50r, 50v

Pile setting and pile pulling, *mic-*   125
tere et extrhare palos.
I 12v, 46v, 47v, 54v, 55r; IV 62r;
S 100r, 116v, 117r, 135v; DM 80r,
80v

Diving equipment.   38 f.
I 12r, 30r, 31r, 57r; III 44r (bottom
right)

Flotation equipment.   44
II 90v, 91r (also FG); DM 78r, 78v

Harbors, *portus,* and harbor   34 ff.
works (also river works), bridge
works.
I 13r, 49r; S 107v, 108v

Dredging equipment.   *
I 5r, 46r, 69v (perpetual motion?);
S 104v

Ship-loading equipment.   61 f.
I 1r, 4r; III 40v–41r (also FG),
41v, 43v; SF 23v

Worm and worm gear   192 f.
GC 45, 57, 89, 90, 91, 92, 93, 107,
141, 147, 171, 191, 193, 194, 197,
198, 199, 200, 202, 205, 206, 209,
210, 211, 213, 219; MC 72

Perpetual motion.   192
I 58r

3. WATER SUPPLY
De aquis actingendis

Aqueductus; underground water   55
main, *butinus,* with risers.
I 41v, 51r, 52r, 52v, 64v, 71v; III
28r, 32v–33r, 34v–35r, 48v–49r;
S 107v, 112r, 112v, 113r, 113v–
114r, 115v, 117v, 118v; DM 50v–
51r

Archimedean Screw (shown ver-   153
tical).
I 38v

Bellows, *mantachi,* with various   37
types of actuation, also as air
blowers.
I 30v, 31v, 53r, 53v, 61v, 63r; DM
43v

Chain pump, *edifitium cum pen-*   51 f.
dentibus ad istar fungorum sive
trochorum.
II 80v; III 30r

Piston pump, *canne agentes et*   49 ff.
patientes, canna cum lingno et ani-
mellis, usually actuated by levers
or cam shafts, sometimes by
crankshafts.
I 42r, 42v, 43r, 53r, 53v, 54r
(partly erased), 58v, 60v, 62v, 63v,
72r; II 82r, 85r, 88r; III 29r, 33v–34r,
44v–45r; S 105v; DM 38r, 38v,
39r, 40r

Siphon, *canna ascendens et de-*   53 ff.
scendens.
I 55v, 60r, 63r, 68r, 70r, 73v;

II 83r, 94v; III 32r, 56v–57r (copied,
MC 111; copied with modifications, MC 77); S 105r, 115r

Water buckets, *situle, vasa, urcei,*   36 f.
on ropes or chains or wheels.
1 15r, 15v, 19r, 19v, 20v, 35v, 36r,
39r, 51v, 59r; II 87r, 96v; III 30r,
39r (also FG); V 64r; DM 36r,
36v, 37r, 41v, 43r

Water scoops on levers.   *
I 15v, 56r, 56v, 71r; II 92r, 96r; S
105r; DM 37v, 38r, 39v, 40v, 41r

4. MILLS
De molendinis
including grist mills, sawmills,
and power generators for water
supply means and other equipment.

Driven by animal power or man   58 f.
power.
I 65r, 65v, 66v; III 36r ff; DM 32v

Treadwheel, *rota magna.*   52 f.
I 66r (Copied, MC 70); III 31r

Driven by waterwheel on a hori-   66
zontal shaft, *reticinum.*
I 36r, 40r, 74r; III 34v–35r; 44v–
45r, 56r; IV 66r; S 97r, 103r, 103v,
106r; MC 59, 62, 68, 71, 72, 73, 79,
86, 88, 102, 165 with horizontal-
shaft windmill; 62 and 72 have
flywheels on vertical shafts driving the millstones

Driven by waterwheel, wind   126
motor, *giratum a vento,* or other
motor on a vertical drive shaft.
I 65v, 66v; II 83v, 87r; III 33v–34r;
IV 64r; MC 60, 61, 63, 64, 66, 67
(all these in MC have flywheels,
and such wheels are also shown
on drives using hand crank, 65,
or animal actuation, 76); 69, 74,

12r, 13r, 13v, 14v, 20r; S 101r,
102v, 116r

Pontoon bridges, *pontes.*   *
I 19v, 43v; II 89v; S 107r, 118v;
DM 15v

6. ARCHITECTURAL DRAWINGS

Battlements.   *
II 80r

Chimneys.   37
I 32r to 34r, 34v, 35r

Fountains.   49
III 28r

Door hinges.   *
I 18r

Columns.   61 f.
III 40v–41r

Roofs.   128
I 37r, 45r, 48r, 48v; S 120v, 121r;
IV 74r, 74v, 75r

Stairs, *scale.*   *
I 73v, 74v

Steam blowers, probably for   *
chimneys.
I 69r

Tools.   37
I 16v, 18r; III 40v–41r; IV 62r, 62v

INDEX